Additive Manufacturing

T0187921

Manufacturing Design and Technology Series

Series Editor
J. Paulo Davim

Additive Manufacturing
Applications and Innovations

Edited by
Rupinder Singh
J. Paulo Davim

CRC Press
Taylor & Francis Group
Boca Raton London New York

CRC Press is an imprint of the
Taylor & Francis Group, an **informa** business

CRC Press
Taylor & Francis Group
6000 Broken Sound Parkway NW, Suite 300
Boca Raton, FL 33487-2742

First issued in paperback 2021

ISBN-13: 978-0-367-78094-4 (pbk)
ISBN-13: 978-1-138-05060-0 (hbk)

Library of Congress Cataloging-in-Publication Data

Names: Singh, Rupinder, 1979- editor. | Davim, J. Paulo, editor.
Title: Additive manufacturing : applications and innovations / [edited by] Rupinder Singh and J. Paulo Davim.
Other titles: Additive manufacturing (CRC Press : 2018)
Description: First edition. | Boca Raton, FL : CRC Press/Taylor & Francis Group, 2018. | Includes bibliographical references and index.
Identifiers: LCCN 2018027130 | ISBN 9781138050600 (hardback : acid-free paper) | ISBN 9781315168678 (ebook)
Subjects: LCSH: Three-dimensional printing. | Manufacturing processes.
Classification: LCC TS171.95 .A3322 2018 | DDC 670--dc23
LC record available at https://lccn.loc.gov/2018027130

Visit the Taylor & Francis Web site at
http://www.taylorandfrancis.com

and the CRC Press Web site at
http://www.crcpress.com

Contents

Preface

Today, additive manufacturing (AM) technologies are regarded as a revolutionary manufacturing technology. In this present commercial world, as the service life of components is relatively reduced due to frequent changes in the customer demand as well as product design, these innovative technologies have made significant strides for the past few years with immense relevance to design creativity, digital fabrication, and time-compression with varieties of product. As per definition of AM, these processes involve the fabrication of three-dimensional parts through material addition in the form of thin layers. AM technologies are being widely used for the fabrication of functional/non-functional parts with a variety of materials such as polymers, metals, ceramics and thereby combinations. The parts thus fabricated could be used for various end-user applications such as architecture, construction, industrial design, automotive, aerospace, military, engineering, dental and medical industries, biotech (human tissue replacement), fashion, footwear, jewellery, eyewear, education, geographic information systems, food and many other fields. This book highlights the latest innovations in the area of AM along with basic fundamentals, mathematical modelling, suitable numerical problems and future scope/directions, especially for undergraduate and postgraduate students.

In spite of the best of our efforts, there are bound to be some mistakes. The same may kindly be brought to our notice for rectification. We sincerely hope that *Additive Manufacturing: Applications and Innovations*, which is of the Manufacturing Design and Technology Series meets the expectations of readers on this vital subject of production technology.

Editors

Rupinder Singh received his PhD degree in mechanical engineering in 2006 from Thapar Institute of Engineering and Technology, Patiala (India), the MTech degree in production engineering in 2001, and the BTech production engineering degree in 1999 from the Punjab Technical University, Jalandhar (India). He is a chartered engineer with the Institution of Engineers (India). Currently, he is Professor of the Department of Production Engineering at the Guru Nanak Dev Engineering College, Ludhiana (India). He has more than 17 years of teaching and research experience in manufacturing and production engineering with special emphasis in additive manufacturing. Dr. Singh's interests are rapid casting, nontraditional machining and maintenance engineering. He has guided large numbers of PhD and masters students as well as coordinated and participated in 18 research projects. Dr. Singh has received several scientific and research awards. He is the editor-in-chief of 2 international journals, guest editor of 4 journals, book editor and scientific advisor to 2 international journals. Presently, he is an editorial board member of 5 international journals and acts as reviewer for more than 10 prestigious Web of Science journals. In addition, Dr. Singh authored more than 20 books, 50 book chapters and 424 articles in journals and conferences (h-index 27+/2700+ citations).

J. Paulo Davim received his PhD degree in mechanical engineering in 1997, the MSc degree in mechanical engineering (materials and manufacturing processes) in 1991, the mechanical engineering degree (5 years) in 1986 from the University of Porto (FEUP), the aggregate title (Full Habilitation) from the University of Coimbra in 2005, and the DSc degree from London Metropolitan University in 2013. He has been named Eur Ing by FEANI-Brussels and Senior Chartered Engineer by the Portuguese Institution of Engineers with an MBA and Specialist title in Engineering and Industrial Management. Currently, he is Professor at the Department of Mechanical Engineering of the University of Aveiro, Portugal. He has more than 30 years of teaching and research experience in manufacturing, materials and mechanical engineering with special emphasis in machining and tribology. Dr. Davim interests lie in management and industrial engineering and higher education for sustainability and engineering education. He has guided large numbers of postdoc, PhD, and masters students as well as coordinated and participated in several research projects. Dr. Davim has received several scientific awards. He has worked as evaluator of projects for international research agencies as well as examiner of PhD theses for many universities. He is the editor-in-chief of several international journals, guest editor of journals, book editor, book series editor and scientific advisor for many international journals and conferences. Presently, Dr. Davim is an editorial board member on 25 international journals and acts as a reviewer for more than 80 prestigious Web of Science journals. In addition, he is the editor (and co-editor) of more than 100 books and author (and co-author) of more than 10 books, 80 book chapters, and 400 articles in journals and conferences (more than 200 articles in journals indexed in Web of Science core collection/h-index 45+/6000+ citations and SCOPUS/h-index 52+/8000+ citations).

Contributors

I. P. S. Ahuja
Department of Mechanical
    Engineering
Punjabi University
Patiala, India

K. S. Bindra
Laser Development and Industrial
    Applications Division
Raja Ramanna Centre for
    Advanced Technology
Indore, India

and

Homi Bhabha National Institute
Mumbai, India

Kamaljit Singh Boparai
Mechanical Engineering
    Department
Maharaja Ranjit Singh Punjab
    Technical University
Bathinda, India

Jasgurpreet Singh Chohan
Mechanical Engineering
    Department
Chandigarh University
Mohali, India

A. N. Jinoop
Laser Development and Industrial
    Applications Division
Raja Ramanna Centre for
    Advanced Technology
Indore, India

and

Homi Bhabha National Institute
Mumbai, India

Ranvijay Kumar
Department of Production
    Engineering
Guru Nanak Dev Engineering
    College
Ludhiana, India

and

Department of Mechanical
    Engineering
Punjabi University
Patiala, India

S. Maidin
Faculty of Manufacturing
    Engineering
Universiti Teknikal Malaysia
    Melaka
Durian Tunggal, Malaysia

M. K. Muhamad
Faculty of Manufacturing
    Engineering
Universiti Teknikal Malaysia
    Melaka
Durian Tunggal, Malaysia

Pulak M. Pandey
Department of Mechanical
    Engineering
IIT Delhi
New Delhi, India

C. P. Paul
Laser Development and Industrial
    Applications Division
Raja Ramanna Centre for
    Advanced Technology
Indore, India

and

Homi Bhabha National Institute
Mumbai, India

E. Pei
Department of Design
College of Engineering and
    Physical Sciences
Brunel University London
Uxbridge, United Kingdom

Jatender Pal Singh
Director General of Quality
    Assurance Department of
    Defense Production, DGQA
New Delhi, India

Narinder Singh
Mechanical Engineering
    Department
Guru Nanak Dev Engineering
    College
Ludhiana, India

Rupinder Singh
Production Engineering
    Department
Guru Nanak Dev Engineering
    College
Ludhiana, India

Sunpreet Singh
School of Mechanical Engineering
Lovely Professional University
Phagwara, India

chapter one

Developing functionally graded materials by fused deposition modelling assisted by investment casting

Sunpreet Singh and Rupinder Singh

Contents

1.1 Functionally graded material

Because of the mechanical progression, there is a huge change in the assembling area. In response to the previous two decades, the globe is requesting material that can manage the extraordinary ecological conditions. Especially in aviation and vehicle divisions, the necessity is for a sort of material that has high quality, durability, hardness, delayed administration life and so forth. Metal network composites initially rose as a unique innovation with enhanced execution – right off the bat as a propelled military framework gave an essential inspiration to materials improvement [1].

Metal composites are broadly utilized as a part of aviation and the automotive industry because of their improved properties – for example, versatile modulus, hardness, elasticity and incredible wear resistance in mix with lightweight materials when contrasted with unreinforced partners [2–6]. These are designed as a blend of metal framework and fortification materials. They have hard and hardened strengthening stages in metallic lattice [7]. From past decades, inquiries have been done for the advancement of lightweight composites fortified with strands, hair and particulate. Among metallic grid materials, Al, Mg, Ti, Cu and their combinations are profoundly utilized on account of the upsides of their formability and cost [8]. A gathering of new composite materials is fundamentally considered for upgraded designing administrations with aluminium metal network composites [9]. Furthermore, because of high solidness, unrivalled mechanical properties and oxidation resistance of alumina, a unique class of composite materials is produced in terms of Al-alumina composites [10,11]. The heterogeneous material class with gradual variation in properties is collectively referred to as functionally graded material. Similar to many other man-made materials, functionally graded natural materials such as bamboo have been used for thousands of years in decorating and construction [12]. In 1972, the theoretical implications of these materials were investigated by Bever et al. [13,14]. That time, the manufacturing systems were limited and rigid, which delayed the development of graded structure materials [15]. Japan is the first country that introduced the scientific term of functionally graded material in 1984 [16]. These functionally graded materials were initially used as thermal barrier materials for aerospace structural applications and fusion reactors but now are considered a potential structural material candidate for high-speed automotive and aerospace vehicles [17]. In the past few years, some researchers began to pay attention to the contact problem of functionally graded materials. There are numerous techniques (Figure 1.1) for the development of functionally graded material that involve the sol-gel process, super plastic forming with diffusion bonding and additive manufacturing.

In the sol-gel process, a fibre stacking and chemical method uses additives, chemicals and hot pressing to fabricate graded ceramic matrix composites in an innovative way. Super plastic forming with diffusion bonding uses a plastic metal or ceramic as basic material and is helpful in reducing the temperature and pressure required due to a low strain rate. Moreover, this technique provides flexibility in design, reduces mechanical joints and is used primarily for manufacturing aircraft parts from titanium alloys. Additive manufacturing processes are helpful in achieving the complex functionally graded material geometry with composition change (gradient in structure, porosity and density), structural change or porosity change [18]. There are many additive manufacturing systems for the processing of functionally graded material and among them all, fused

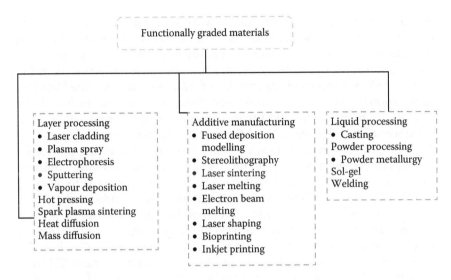

Figure 1.1 Various methods of functionally graded material fabrication.

deposition modelling, selective laser sintering, laser engineered net shaping, selective laser melting and electron beam melting are presenting best technologies in the class.

Additive manufacturing technologies are known for their various benefits as discussed as follows [19]:

- No tooling is needed, which significantly reduces production ramp-up time and expense.
- Small production batches are feasible and economical.
- There is a possibility to quickly change design.
- The product can be optimized for function (e.g. optimized cooling channels).
- Economical custom products (batch of one) can be created.
- Waste reduction is possible.
- There is a potential for simpler supply chains – shorter lead times, lower inventories.
- Design customization is available.

Additive manufacturing is an advanced innovation for creating physical articles layer by layer from a three-dimensional CAD document. The procedure starts with creating a three-dimensional CAD model of the question with every one of its points of interest and measurements. This three-dimensional CAD record can likewise be produced from MRI/CT information that is examined through scans. This three-dimensional CAD document is cut or segmented into various thin two-dimensional areas by a PC program.

A short time later, the various cuts of two-measurement areas are sent to the printing machine, which sets it down on a stage consistently [20,21]. The procedure may take from a couple of hours to a couple of days to deliver a part, contingent upon size, geometry and exactness. Additive manufacturing (AM) advancements are predominantly needed upon the Standard Tessellation Language (.STL) record concerning each part, and the initial step is to get the required .STL document. In Kalita et al. [22], researchers fabricated scaffolds with different complex internal architectures and controlled porosity (36%, 48% and 52%) with a fused deposition modelling process by using polypropylene polymer and tricalcium phosphate ceramic. Similarly, Chung and Das [23] studied a selective laser sintering process to fabricate functionally graded polymer nanocomposites of nylon-11 filled with 0%–10% vol. of 15 nm silica nanoparticles. In one of the other studies [24], a selective laser sintering technology was used for the development of functionally graded scaffolds with a stiffness gradient that mimics the native bone by controlling their structure and porosity. Mumtaz and Hopkinson [25] studied selective laser melting for functionally graded specimens of super nickel alloy and ceramic compositions with discrete changes between layered compositions. Research by Hazlehurst et al. [26] confirmed that selective laser melting can repeatedly manufacture functionally graded material 48% lighter and 60% more flexible than a traditional fully dense femoral implant. Zhang et al. [27] fabricated functionally graded material with composition change from pure titanium to 40% vol. of titanium carbide with a laser engineered net shaping process by adjustment of process parameters and real-time variation of the material feeding ratio during the process. Durejko et al. [28] demonstrated the fabrication of thin wall tubes Fe_3Al/SS-316L material with composition change perpendicularly to the wall of the tube. In a study on electron beam melting [29], titanium/molybdenum functionally graded material with high temperature resistance was fabricated with fine appearance and good interface. Similarly, this setup was also studied by Parthasarathy et al. [20] to fabricate functionally graded material with periodic cellular structures specifically targeted for biomedical applications. The reported functionally graded material fabricated by additive manufacturing showed the potential of the technique besides the existence of challenges in additive manufacturing.

The minimum wall thickness and internal geometries with very fine structures below 1 mm are difficult to obtain with additive manufacturing. Moreover, there are issues as the limited variety of workhorse material and waste associated due to support structures. Recently, researchers have developed Al-alumina (Al-Al_2O_3) graded composites, by using an alternative reinforced fused deposition modelling sacrificial pattern in an investment casting process [31–35]. Basically, investment casting bottlenecks such as intrinsic weakness of wax pattern leading to shell cracking, expensive tooling, long production cycle time and difficulties in product

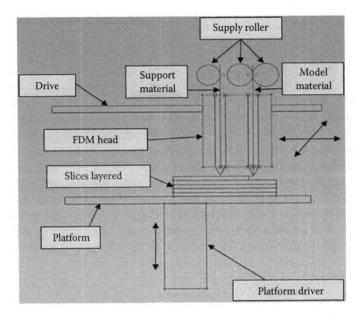

Figure 1.2 Schematic of FDM system. (From Singh, R. and Singh, S., Development of functionally graded material through the fused deposition modelling-assisted investment casting process: A case study, in: S. Hashmi (Ed.), *Reference Module in Materials Science and Materials Engineering*, Elsevier, Oxford, UK, 2016 [38].)

development have been successfully overcome by fused deposition modelling assistance [36,37]. Initially, new composite material proportions were selected and used to fabricate reinforced fused deposition modelling (Figure 1.2) filament on a single screw extruder.

This alternatively prepared filament was used to fabricate investment casting sacrificial patterns on an existing fused deposition modelling uPrint-SE© system, available at Manufacturing Research Lab, GNDEC, Ludhiana (India).

Patterns were finished for improvement of their surface finish using a barrel finishing process. Moulds were prepared in a similar way as conventional investment casting process and casting were carried out using Al-6063 alloy. Process parameters like filament proportion (A), volume of pattern (B), density of fused deposition modelling pattern (C), barrel finishing cycle time (D), barrel finishing media weight (E) and number of slurry layers (F) were already investigated for their effect on mechanical, tribological and metallurgical properties of composite developed by Singh and Singh of MRL, GNDEC (Ludhiana) [32–35]. Efforts in the present chapter are aimed at developing mathematical models for hardness, wear, dimensional accuracy and surface roughness of composites by using Buckingham's π-theorem.

1.2   A case study of an alternative method of developing functionally graded material

The procedure adopted in the present research work is shown in Figure 1.3. Initially, the melt flow index was performed (as per the ASTM-D-1238 standard) [39] at 230°C and 3.8 kg weight in order to select the appropriate proportions of alternative reinforced filament ingredients that should match the rheology of commercially used ABS-P400 material (for uPrintSE setup). Then, a single screw extruder (L/D ratio 20) was used for preparation of alternative reinforced fused deposition modelling filaments of 60%nylon6/30%Al/10%Al$_2$O$_3$ (c$_1$) and 60%nylon6/28%Al/12%Al$_2$O$_3$ (c$_2$) proportions. The alternative filaments developed were tested mechanically, and tensile strength of c$_1$ and c$_2$ proportion was found as 21.65 and 21.53 N/mm^2, respectively. The filaments developed were then fed into a fused deposition system to fabricate investment casting sacrificial patterns in cubical shape having volume as 17,576, 27,000 and 39,304 mm^3 at low, high and solid density of the uPrintSE system. However, any shape and size can be selected for the sacrificial pattern of

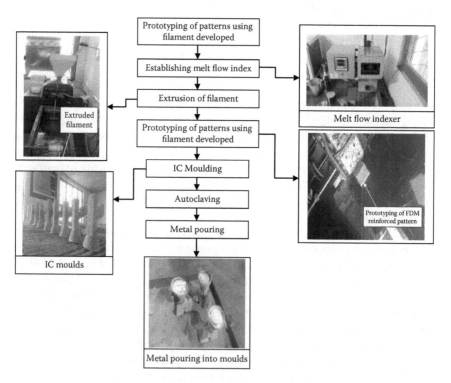

Figure 1.3 Procedure of functionally graded material development.

investment casting. Barrel finishing was also performed to improve the finish of reinforced patterns. Investment casting moulds were prepared by coating casting trees with refractory layers of silica. Autoclaving and baking were performed simultaneously at 1150°C (keeping the pouring cup upright in order to lock the abrasive Al_2O_3 particles in the mould cavity). A molten Al-6063 matrix was then poured into the resulting cavity.

Taguchi L18 array has been selected to design experimentation (as shown in Table 1.1). Castings produced were tested for hardness, wear, dimensional accuracy and surface roughness using HVS-1000BVM hardness tester (HV0.01 scale), pin-on-disc, Vernier calliper and Mitutoyo SJ-210 surface roughness tester, respectively [32–35]. Results are given in Table 1.2. The percentage contribution (given in Table 1.3) of input process parameters in hardness, wear, dimensional accuracy and surface roughness was calculated through statistical software Minitab-17.

It has been found from Table 1.3 that in cases of hardness, wear, dimensional accuracy and surface roughness, the most contributing process parameters are density of pattern, volume of pattern, barrel finishing media weight and volume of pattern, respectively. For further justification of methodology, Figure 1.4 shows the scanning electron microscopic

Table 1.1 Taguchi L18-based control log of experimentation

		Design of experimentation				
Exp. no.	A, N/mm²	B (mm³)	C (×10⁻⁶), N/mm³	D (sec.)	E (N)	F, mm
1	21.65	17,576	5.12	1200	98	11.5
2	21.65	17,576	7.63	2400	147	13
3	21.65	17,576	9.163	3600	196	15
4	21.65	27,000	5.12	1200	147	13
5	21.65	27,000	7.63	2400	196	15
6	21.65	27,000	9.163	3600	98	11.5
7	21.65	39,304	5.12	2400	98	15
8	21.65	39,304	7.63	3600	147	11.5
9	21.65	39,304	9.163	1200	196	13
10	21.53	17,576	5.12	3600	196	13
11	21.53	17,576	7.63	1200	98	15
12	21.53	17,576	9.163	2400	147	11.5
13	21.53	27,000	5.12	2400	196	11.5
14	21.53	27,000	7.63	3600	98	13
15	21.53	27,000	9.163	1200	147	15
16	21.53	39,304	5.12	3600	147	15
17	21.53	39,304	7.63 × 10⁻⁶	1200	196	11.5
18	21.53	39,304	9.163	2400	98	13

Table 1.2 Results obtained from control log of experimentation

Exp. no.	Hardness[a] H (1 HV = 9.807 N/mm²)	Wear[b] (mm) W	Dimensional accuracy[b] (mm) Δd[a]	Surface roughness[b] (mm) R_a
1	878.2893	0.128	0.026	0.00475
2	900.558	0.106	0.033	0.00515
3	1128.739	0.087	0.02	0.00477
4	788.0373	0.171	0.056	0.00436
5	848.8593	0.167	0.063	0.00558
6	1128.15	0.135	0.053	0.00609
7	756.351	0.182	0.043	0.00536
8	901.539	0.168	0.08	0.00565
9	985.5126	0.155	0.016	0.00639
10	916.8426	0.122	0.016	0.0047
11	941.3676	0.109	0.076	0.00456
12	1317.777	0.095	0.056	0.00465
13	935.1873	0.153	0.033	0.00529
14	920.178	0.140	0.05	0.00588
15	1025.734	0.124	0.06	0.00584
16	825.3153	0.153	0.033	0.00856
17	1005.133	0.136	0.043	0.00571
18	1042.116	0.125	0.046	0.00588

[a] Analyzed for 'higher the better' condition.
[b] Analyzed for 'smaller the better' condition.

Table 1.3 Percentage contribution of input process parameters

Source	Hardness (%)	Wear (%)	Dimensional accuracy (%)	Surface roughness (%)
Proportion of filament	7.69	6	0.76	4.16
Volume of fused deposition modelling reinforced pattern (mm³)	8.85	62[a]	16.95	43.84[a]
Density of fused deposition modelling pattern	65.75[a]	23.9	19.83	3.03
BF time (mint)	1.03	0.40	3.30	6.45
BF media weight (kg)	0.8	0.080	31.71[a]	2.94
Number of IC slurry layers	14.14	0.11	8.97	5.72
Residual error	1.74	8.03	18	33.86
Total	100	100	100	100

[a] Highly contributed factor.

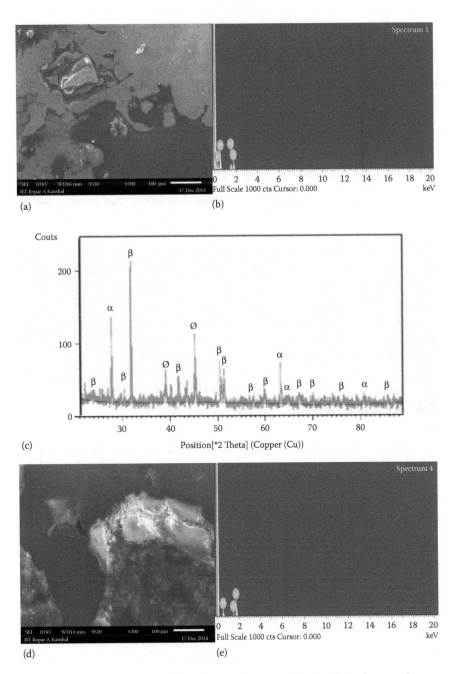

Figure 1.4 Characterization of castings; micrograph (a, d and g), electron dispersion spectroscopy (b, d and h) and X-ray diffraction analysis (c, f and i) for experiments #16, #17 and #18. *(Continued)*

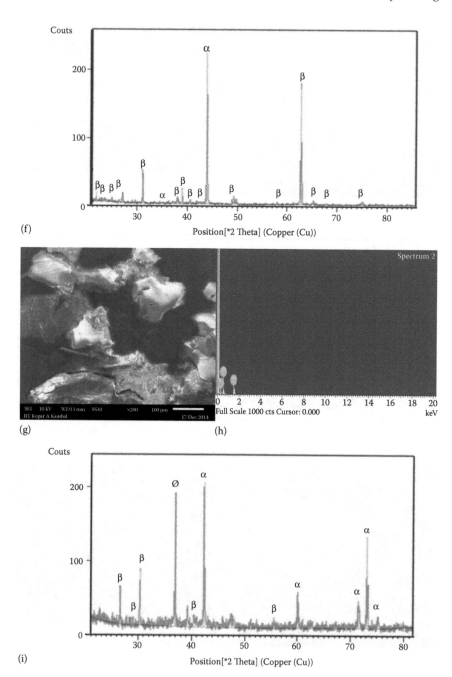

Figure 1.4 **(Continued)** Characterization of castings; micrograph (a, d and g), electron dispersion spectroscopy (b, d and h) and X-ray diffraction analysis (c, f and i) for experiments #16, #17 and #18.

analysis along with electron dispersion spectroscopy and X-ray diffraction analysis has been made on experiments #16, #17 and #18 of Table 1.1.

It has been observed from micrographs that Al_2O_3 particles are presented in aluminium matrix, and this has been cross-verified through the peaks of Al and O and Al_2O_3 in electron dispersion spectroscopy and X-ray diffraction plots, respectively. Further, the distribution of Al_2O_3 particles in the castings is in order of #18 > #17 > #16.

1.3   Development of mathematical modelling for functionally graded materials using Buckingham π-theorem

The dimension-less analysis has proven as one of the most successful and effective approaches for generating analytic equations in complex systems (with larger number of variables) [40]. Today, dimensionless analysis is the concept with which influence of variables can be reduced by means of some physical equations [41]. By using the Vashy-Buckingham π-theorem it is possible to predict hardness, wear, dimensional accuracy and surface roughness of composites with a minimal set of dimensional combinations. The approach illustrates that if a physical problem contains 'n' parameters (input and output), where 'm' represents the dimensions of the quantities (M, L, T and ϑ), then the subtraction of both – that is 'n – m' – gives the number of independent parameters (i.e. πs) to assume. The equations are derived for assumed independent parameters and further used for the development of dimensional relationships among them [42]. Since the output parameters in the present study are hardness, dimensional accuracy and surface roughness and are dependent upon selected input parameters, the basic dimensions can be assumed as:

- M (mass)
- L (length)
- T (time)
- ϑ (temperature)

And the dimensions for the input parameters would be:

1. Hardness (H) as $ML^{-1}T^{-2}$
2. Wear (W) as L
3. Dimensional accuracy as L
4. Surface roughness as L
5. Proportion of filament (P) in terms of tensile strength of the filament, $ML^{-1}T^{-2}$
6. Volume of reinforced pattern (V) as L^3

   7. Density of pattern (ρ) as ML^{-3}
   8. Barrel finishing time (t) as T
   9. Barrel finishing media weight (W) as M
   10. Number of slurry layers (l) as L

1.3.1   Wear modelling

In the present study, hardness as the first output parameter is kept as a function of all input process parameters as given in Equation 1.1.
   Now,

$$W = f(P, V, \rho, t, m, l) \tag{1.1}$$

In the present case 'n' is 7 and 'm' is 3. Therefore, the number of independent quantities will be 'n − m' – that is, 4 (7 − 3 = 4). Note that π_1, π_2, π_3 and π_4 are four-dimensional quantities. Based on Table 1.3, barrel finish time, barrel finish media weight and number of investment casting slurry layers are the least significant in the present case so they will directly go in the 'π' groups.

$$\pi_1 = W(t)^{a_1} (m)^{b_1} (l)^{c_1} \tag{1.2}$$

$$\pi_2 = V(t)^{a_2} (m)^{b_2} (l)^{c_2} \tag{1.3}$$

$$\pi_3 = \rho(t)^{a_3} (m)^{b_3} (l)^{c_3} \tag{1.4}$$

$$\pi_4 = P(t)^{a_4} (m)^{b_4} (l)^{c_4} \tag{1.5}$$

Now, substituting the dimensions of each quantity in Equations 1.2 through 1.5 and equating them to zero, the ultimate exponent of each basic dimension has been achieved, since the 'π's' are dimensional groups. Thus, a_i, b_i and c_i (where i = 1, 2, 3...) can be solved.

$$\pi_1 = L(T)^{a_1} (M)^{b_1} (L)^{c_1} \tag{1.6}$$

Here,

$$M: b_1 = 0$$

$$L: 1 + c_1 = 0$$

$$T: a_1 = 0$$

So,

$$a_1 = 0, b_1 = 0 \text{ and } c_1 = -1$$

Thus,

$$\pi_1 = m/l \tag{1.7}$$

Similarly, we get;

$$\pi_2 = L^3 (T)^{a_2} (M)^{b_2} (L)^{c_2} \tag{1.8}$$

Here,

$$M: b_2 = 0$$

$$L: 3 + c_2 = 0$$

$$T: a_2 = 0$$

So,

$$a_2 = 0, b_2 = 0 \text{ and } 3 + c_2 = 0, \text{ so } c_2 = -3$$

Thus,

$$\pi_2 = v/l^3 \tag{1.9}$$

Similarly, we get;

$$\pi_3 = ML^{-3} (T)^{a_3} (M)^{b_3} (L)^{c_3} \tag{1.10}$$

Here,

$$M: 1 + b_3 = 0$$

$$L: -3 + c_3 = 0$$

$$T: a_3 = 0$$

So,

$$a_3 = 0, b_3 = -1 \text{ and } -3 + c_3 = 0, \text{ so } c_3 = 3$$

Thus,

$$\pi_3 = \rho l^3/m \tag{1.11}$$

Similarly, we get;

$$\pi_4 = ML^{-1}T^{-2}(T)^{a_4}(M)^{b_4}(L)^{c_4} \tag{1.12}$$

Here,

$$M: 1 + b_4 = 0$$

$$L: -1 + c_3 = 0$$

$$T: -2 + a_4 = 0$$

So,

$$a_4 = 2,\ b_4 = -1 \text{ and } c_4 = 1$$

Thus,

$$\pi_4 = Pt^2 l/m \tag{1.13}$$

The ultimate relationship can be assumed to be of the form;

$$\pi_1 = f(\pi_2, \pi_3 \text{ and } \pi_4)$$

Or,

$$W/L = f\left(\frac{V}{l^s}, \frac{\rho l^3}{m} \text{ and } \frac{Pt^2 l}{m}\right)$$

Or,

$$W = K.V.\rho.P.t^2.l/m^2 \tag{1.14}$$

where 'K' is constant of proportionality.

It was experimentally observed that 'W' directly goes with 'V', which means that volume of pattern has strongly affected the wear of the composites. Therefore, 'V' was taken as representative for development of a mathematical equation. The actual experimental data for 'V' have been collected and plotted in Figure 1.5. The data collected were further used to find the best fitted curve. A second-order polynomial equation was determined with regression equals to 1.

So,

$$W = \left(-2 \times 10^{-10} V^2 + 1 \times 10^{-05} V - 0.03\right) \rho.P.t^2.l/m^2 \tag{1.15}$$

- Corollary

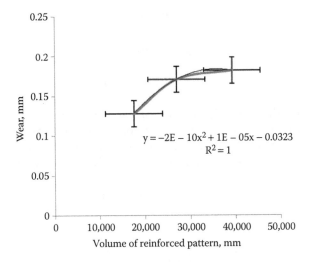

Figure 1.5 Wear versus volume of fused deposition modelling pattern plot.

Equation 1.15 was further used for testing and for verification of a mathematical equation with experiment #3 (optimum value of wear from Table 1.1). The experimental value for 'W' in experiment #3 is 0.087 mm. By putting the input parameter values of experiment #3 (as per Table 1.1) in Equation 1.15:

Result of wear comes out as $W = 0.084$ mm (predicted value)

The result of wear obtained using a mathematical equation agrees well with the experimental value (see third row of Table 1.1). The model error (i.e. 3.44%) is below the permissible limit of 5%, justifying toward the validity of the model.

1.3.2 Hardness modelling

Hardness as the first output parameter is kept as a function of all input process parameters as given in Equation 1.16.
So,

$$H = f(P, V, \rho, t, m, l) \qquad (1.16)$$

In the present case, 'n' is 7 and 'm' is 3. Therefore, the number of independent quantities will be 'n – m' – that is, 4 (7 – 3 = 4). Note that π_1, π_2, π_3 and π_4 are four-dimensional quantities. Based on Table 1.3, proportion of filament, barrel finishing time and barrel finishing media weight are the least significant in the present case so they will directly go in the 'π' groups.

$$\pi_1 = H(P)^{a_1}(t)^{b_1}(m)^{c_1} \qquad (1.17)$$

$$\pi_2 = \rho\,(P)^{a_2}\,(t)^{b_2}\,(m)^{c_2} \tag{1.18}$$

$$\pi_3 = l\,(P)^{a_3}\,(t)^{b_3}\,(m)^{c_3} \tag{1.19}$$

$$\pi_4 = V\,(P)^{a_4}\,(t)^{b_4}\,(m)^{c_4} \tag{1.20}$$

Now, substituting the dimensions of each quantity in Equations 1.17 through 1.20 and equating them to zero, the ultimate exponent of each basic dimension has been achieved, since the 'π's' are dimensional groups. Thus, a_i, b_i and c_i (where i = 1, 2, 3...) can be solved. Solving π_1:

$$\pi_1 = ML^{-1}T^{-2}\left(ML^{-1}T^{-2}\right)^{a_1}(T)^{b_1}(M)^{c_1} \tag{1.21}$$

Here,

$$M: 1 + a_1 + c_1 = 0$$

$$L: -1 - a_1 = 0$$

$$T: -2 - 2a_1 + b_1 = 0$$

So,

$$a_1 = -1,\ b_1 = 0 \text{ and } c_1 = 0$$

Thus,

$$\pi_1 = H/F \tag{1.22}$$

Similarly, we get;

$$\pi_2 = ML^{-3}(ML^{-1}T^{-2})^{a_2}(T)^{b_2}(M)^{c_2} \tag{1.23}$$

Here,

$$M: 1 + a_2 + c_2 = 0$$

$$L: -3 - a_2 = 0$$

$$T: -2a_2 + b_2 = 0$$

So,

$$a_2 = -3,\ b_2 = -6 \text{ and } c_2 = 2$$

Thus,

$$\pi_2 = \rho/F^3 t^6 \tag{1.24}$$

Similarly, we get;

$$\pi_3 = L\,(ML^{-1}T^{-2})^{a_3}\,(T)^{b_3}\,(M)^{c_3} \tag{1.25}$$

Here,

$$M:\ a_3 + b_3 = 0$$

$$L:\ 1 - a_3 = 0$$

$$T:\ -2a_3 + b_3 = 0$$

So,

$$a_3 = 1,\ b_3 = 2 \text{ and } c_3 = -1$$

Thus,

$$\pi_3 = lFT^2/m \tag{1.26}$$

Similarly, we get;

$$\pi_4 = L^3(ML^{-1}T^{-2})^{a_4}\,(T)^{b_4}\,(M)^{c_4} \tag{1.27}$$

Here,

$$M:\ a_4 + c_4 = 0$$

$$L:\ 3 - a_4 = 0$$

$$T:\ 0 - 2a_4 + b_4 = 0$$

So,

$$a_4 = 3,\ b_4 = 6 \text{ and } c_4 = -3$$

Thus,

$$\pi_4 = VF^3 t^6/m^3 \tag{1.28}$$

The ultimate relationship can be assumed to be of the form;

$$\pi_1 = f\,(\pi_2,\ \pi_3 \text{ and } \pi_4)$$

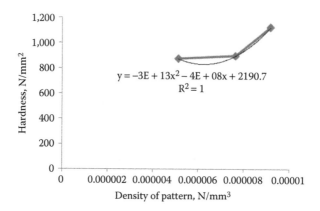

Figure 1.6 Hardness versus density of pattern plot.

Or

$$H/P = f\left(\frac{\rho}{F^3t^6}, \frac{lFt^2}{m} \text{ and } \frac{VP^3t^6}{m^3}\right)$$

Or

$$H = K.\rho.P^2.l.t^2V/m^4 \tag{1.29}$$

'K' is a constant of proportionality.

It has been experimentally found that hardness directly goes with ρ; this means that density of pattern has strongly affected the hardness of casted composites. Therefore, 'ρ' has been taken as representative for development of a mathematical equation. The actual experimental data for 'ρ' have been collected and plotted in Figure 1.6. The data collected have been further used to find the best-fitted curve. A second-order polynomial equation has been determined with regression equal to 1.

The final mathematical model for hardness is:

$$H = \left[(30 + 13(\rho^2) - 40 + 08(\rho) + 2190.7\right]P^2.L.t^2.V/m^4 \tag{1.30}$$

1.3.3 Dimensional accuracy modelling

In the present study, dimensional accuracy as a second output parameter is kept as a function of all input process parameters as given in Equation 1.31.

Now,

$$\Delta d = f(P, V, \rho, t, m, l) \tag{1.31}$$

Based on Table 1.3, filament proportion, barrel finishing time and number of slurry layers are the least significant in the present case so they will directly go in the 'π' groups.

So,

$$\pi_1 = \Delta d \ (P)^{a_1} (t)^{b_1} (L)^{c_1} \tag{1.32}$$

$$\pi_2 = m \ (P)^{a_2} (t)^{b_2} (L)^{c_2} \tag{1.33}$$

$$\pi_3 = \rho(P)^{a_3} (t)^{b_3} (L)^{c_3} \tag{1.34}$$

$$\pi_4 = V \ (P)^{a_4} (t)^{b_4} (L)^{c_4} \tag{1.35}$$

Now, substituting the dimensions of each quantity in Equations 1.32 through 1.35, and equating them to zero, the ultimate exponent of each basic dimension has been achieved, since the 'π's' are dimensional groups. Thus a_i, b_i and c_i (where i = 1, 2, 3...) can be solved. Solving π_1:

$$\pi_1 = L \left(ML^{-1}T^{-2} \right)^{a_1} (T)^{b_1} (L)^{c_1} \tag{1.36}$$

Here,

$$M: a_1 = 0$$

$$L: 1 - a_1 + c_1 = 0$$

$$T: -2a_1 + b_1 = 0$$

So,

$$a_1 = 0, b_1 = 0 \text{ and } c_1 = -1$$

Thus,

$$\pi_1 = \Delta d / l \tag{1.37}$$

Similarly, we get;

$$\pi_2 = M \left(ML^{-1}T^{-2} \right)^{a_2} (T)^{b_2} (L)^{c_2} \tag{1.38}$$

Here,

$$M: 1 + a_2 = 0$$

$$L: -a_2 + c_2 = 0$$

$$T: -2a_2 + b_2 = 0$$

So,

$$a_2 = -1, b_2 = -2 \text{ and } c_2 = -1$$

Thus,

$$\pi_2 = m/Ft^2l \qquad (1.39)$$

Similarly, we get;

$$\pi_3 = ML^{-3}\left(ML^{-1}T^{-2}\right)^{a_3}(T)^{b_3}(L)^{c_3} \qquad (1.40)$$

Here,

$$M: 1 + a_3 = 0$$

$$L: -3 - a_3 + c_3 = 0$$

$$T: -2a_3 + b_3 = 0$$

So,

$$a_3 = -1, b_3 = -2 \text{ and } c_3 = 2$$

Thus,

$$\pi_3 = \rho\, l^2/Ft^2 \qquad (1.41)$$

Similarly, we get;

$$\pi_4 = L^3\left(ML^{-1}T^{-2}\right)^{a_4}(T)^{b_4}(L)^{c_4} \qquad (1.42)$$

Here,

$$M: a_4 = 0$$

$$L: 3 - a_4 + c_4 = 0$$

$$T: -2a_4 + b_4 = 0$$

So,

$$a_4 = 0, b_4 = 0 \text{ and } c_4 = -3$$

Thus,

$$\pi_4 = V/l^3 \tag{1.43}$$

The ultimate relationship can be assumed to be of the form;

$$\pi_1 = f(\pi_2, \pi_3 \text{ and } \pi_4)$$

Or

$$\Delta d/l = f\left(\frac{m}{FP\iota}, \frac{\rho l^2}{Pt^2} \text{ and } \frac{V}{l^3}\right)$$

Or

$$\Delta d = K.\rho.m.V/P^2.l.t^4 \tag{1.44}$$

'K' is a constant of proportionality.

It has been experimentally found that dimensional accuracy is significantly affected by barrel finishing media weight (Table 1.3). Therefore, 'm' has been taken as representative for development of a mathematical equation. The actual experimental data for 'W' have been collected and plotted in Figure 1.7. The data collected have been further used to find the best fitted curve. A second-order polynomial equation has been determined with regression equal to 1.

The final mathematical model for dimensional accuracy is:

$$\Delta d = \left[(-4 \times 10^{-6}\left(m^2\right) + 0.0012(m) - 0.048\right]\rho.V/P^2.l.t^4 \tag{1.45}$$

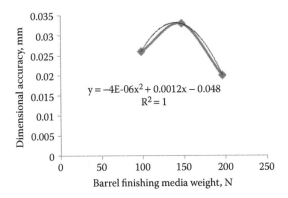

Figure 1.7 Dimensional accuracy versus barrel finishing media weight plot.

1.3.4   *Surface roughness modelling*

In the present study, surface roughness as a third output parameter is kept as a function of all input process parameters as given in Equation 1.46.
Now,

$$R_a = f\left(P, V, \rho, t, m, l\right) \tag{1.46}$$

Based on Table 1.3, barrel finishing media weight, density of pattern and filament proportion are the least significant in the present case so they will directly go in the 'π' groups.
So,

$$\pi_1 = R_a(m)^{a_1}(\rho)^{b_1}(P)^{c_1} \tag{1.47}$$

$$\pi_2 = V(m)^{a_2}(\rho)^{b_2}(P)^{c_2} \tag{1.48}$$

$$\pi_3 = t(m)^{a_3}(\rho)^{b_3}(P)^{c_3} \tag{1.49}$$

$$\pi_4 = l(m)^{a_4}(\rho)^{b_4}(P)^{c_4} \tag{1.50}$$

Now, substituting the dimensions of each quantity in Equations 1.47 and 1.48, and equating them to zero, the ultimate exponent of each basic dimension has been achieved, since the 'π's' are dimensional groups. Thus a_i, b_i and c_i (where i = 1, 2, 3...) can be solved. Solving π_1:

$$\pi_1 = L(m)^{a_1}(ML^{-3})^{b_1}(ML^{-1}T^{-2})\left(ML^{-1}T^{-2}\right)^{c_1} \tag{1.51}$$

Here,

$$M: a_1 + b_1 + c_1 = 0$$

$$L: 1 - 3b_1 - c_1 = 0$$

$$T: -2c_1 = 0$$

So,

$$a_1 = -1/3,\ b_1 = 1/3 \text{ and } c_1 = 0$$

Thus,

$$\pi_1 = R_a(\rho/m)^{1/3} \tag{1.52}$$

Similarly, we get;

$$\pi_2 = L^3(M)^{a_2}(ML^{-3})^{b_2}(ML^{-1}T^{-2})^{c_2} \qquad (1.53)$$

Here,

$$M: a_2 + b_2 + c_2 = 0$$

$$L: 3 - 3b_2 - c_2 = 0$$

$$T: -2c_2 = 0$$

So,

$$a_2 = -1, \, b_2 = 1 \text{ and } c_2 = 0$$

Thus,

$$\pi_2 = V(\rho/m) \qquad (1.54)$$

Similarly, we get;

$$\pi_3 = T(M)^{a_3}(ML^{-3})^{b_3}(ML^{-1}T^{-2})^{c_3} \qquad (1.55)$$

Here,

$$M: a_3 + b_3 + c_3 = 0$$

$$L: -3b_3 - c_3 = 0$$

$$T: 1 - 2c_3 = 0$$

So,

$$a_3 = -1/3, \, b_3 = -1/6 \text{ and } c_3 = 1/2$$

Thus,

$$\pi_3 = t(m)^{-1/3}.(\rho)^{-1/6}.(P)^{-1/3} \qquad (1.56)$$

Similarly, we get;

$$\pi_4 = L(M)^{a_4}(ML^{-3})^{b_4}(ML^{-1}T^{-2})^{c_4} \qquad (1.57)$$

Here,

$$M: a_4 + b_4 + c_4 = 0$$

$$L: 1 - 3b_4 - c_4 = 0$$

$$T: -2c_4 = 0$$

So,

$$a_4 = -1/3, b_4 = 1/3 \text{ and } c_4 = 0$$

Thus,

$$\pi_4 = l(m/\rho)^{1/3} \tag{1.58}$$

The ultimate relationship can be assumed to be of the form;

$$\pi_1 = f\left(\pi_2, \pi_3 \text{ and } \pi_4\right)$$

Or

$$R_a(\rho/m)^{1/3} = f\left(\frac{V\rho}{m}, \frac{t}{(\rho)^{1/6}(Pm)^{1/3}} \text{ and } l\left(\frac{m}{\rho}\right)^{1/3}\right)$$

Or

$$R_a = K.\left(V.t.l.\rho^{1/6}\right)/P^{1/3}.m^{2/3} \tag{1.59}$$

'K' is a constant of proportionality.

It has been experimentally found that surface roughness is significantly affected by volume of pattern (Table 1.3). Therefore, 'V' has been taken as representative for development of a mathematical equation. The actual experimental data for 'V' have been collected and plotted in Figure 1.8. The data collected have been further used to find the best fitted curve. A second-order polynomial equation has been determined with regression equal to 1.

The final mathematical model for surface roughness is:

$$R_a = \left[6 \times 10^{-12}\left(V^2\right) - 3 \times 10^{-7}\left(V\right) - 0.0082\right].\left(t.l.\rho^{1/6}\right)/P^{1/3}.m^{2/3} \tag{1.60}$$

Mathematical modelling on the basis of Buckingham's π-based approach is a very efficient technique for data interpretation obtained after

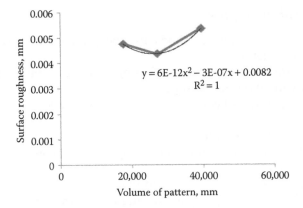

Figure 1.8 Surface roughness versus volume of pattern plot.

conducting experiments as per Taguchi L18 OA. In the present research work, Buckingham's π-theorem was employed for the development of mathematical models of surface roughness, dimensional accuracy, surface hardness and castings obtained using a fused deposition modelling-investment casting process using a reinforced pattern. By using developed mathematical modelling equations, surface roughness, dimensional accuracy, surface hardness and wear of the composite materials may be predicted with minimum experimentations cost/expense.

1.4   Summary

Buckingham's π-based approach is a very efficient technique for data interpretation obtained after conducting experiments as per Taguchi L18 OA. With the help of dimensional analysis, a relationship between the variables influencing a flow problem in terms of dimensional parameters is obtained. This relationship helps in conducting tests on the model. Moreover, the intrinsic value of alternative designs can be predicted with the help of model testing. In the present research work, Buckingham's π-theorem was employed for the development of mathematical models of surface roughness, dimensional accuracy, surface hardness and castings obtained using a fused deposition modelling assisted investment casting process using a reinforced pattern.

The only drawback of this analysis is that the precise functional form among the numbers cannot be obtained, and the coefficient/index obtained cannot be determined by this analysis. However, it can be only determined by experimentation, since the numbers are non-dimensional.

Theoretical questions

Q1. Name the different methods for the production of functionally graded material.

Q2. What are the three benefits of using additive manufacturing technology in investment casting?

Q3. Buckingham's π-theorem can be used for developing _____ models.

Q4. Choose the correct answer:
  i. What does 'M' signify in Buckingham's π-theorem?
   (a) Miles   (b) Meter
   (c) Mass   (d) Model
  ii. What does 'L' signify in Buckingham's π-theorem?
   (a) Litre   (b) Length
   (c) Line   (d) Linear
  iii. What does 'T' signify in Buckingham's π-theorem?
   (a) Tiny molecules   (b) Temperature
   (c) Threshold   (d) Time

Q5. Define 'n' and 'm' in Buckingham's π-theorem with an example.

Q6. Draw the flowchart for the step-by-step illustration of the novel route for functionally graded material development through a fused deposition modelling assisted investment casting process.

Q7. How can the alternative fused deposition filament be developed?

Q8. Name the country and year when the scientific term of functionally graded material was introduced.

Q9. Give examples for the applications of functionally graded materials.

Q10. Is there any restriction over the shape and size of the fused deposition-based sacrificial pattern of investment casting?

Q11. What is the use of barrel finishing process?

Q12. How is percentage contribution calculated?

Q13. What is the formula to decide the number of πs to consider in Buckingham's π-theorem?

Q14. With dimensionless analysis, the effect of variables can be reduced through _____.

Numerical questions (solved)

Q1. What will be the effect on wear of the casted specimen (prepared with a fused deposition modelling-investment casting process), if the volume (V) is reduced to half, density (ρ) is doubled and finishing time (t) is reduced by 40%? Take the suitable value of volume, if required.

Wear follows the following equation:

$$W = \left(-2 \times 10^{-10} V^2 + 1 \times 10^{-5} V - 0.03\right) \rho.P.t^2.l/m^2$$

Solution
By considering the equation of wear,

$$W = \left(-2 \times 10^{-10} V^2 + 1 \times 10^{-5} V - 0.03\right) \rho.P.t^2.l/m^2$$

Now,
In case 1: Let the wear of the resulting casting, volume, density, filament strength, finishing time and media weight be W_1, V_1, ρ_1, P_1, t_1 and m_1.

$$W_1 = \left(-2 \times 10^{-10} V_1^2 + 1 \times 10^{-5} V_1 - 0.03\right) \rho_1.P_1.t_1^2.l/m_1^2$$

In case 2: Let the wear of the resulting casting, volume, density, filament strength, finishing time and media weight be W_2, $0.5V_1$, $2\rho_1$, P_2, t_2 and $0.6m_1$ (as per the given statement).

$$W_2 = \left(-2 \times 10^{-10} \left(0.5 \times V_1\right)^2 + 1 \times 10^{-5} \left(0.5 \times V_1\right) - 0.03\right) 2 \times \rho_1.P_2.t_2^2.l/\left(0.6m_1\right)^2$$

Now, dividing Equation 1.1 by 1.2, as under

$$W_1/W_2 = \left(\left(-2 \times 10^{-10} V_1^2 + 1 \times 10^{-5} V_1 - 0.03\right) \rho_1.P_1.t_1^2.l/m_1^2\right)/$$

$$\left(\left(-2 \times 10^{-10} \left(0.5 \times V_1\right)^2 + 1 \times 10^{-5} \left(0.5 \times V_1\right) - 0.03\right) 2 \times \rho_1.P_2.t_2^2.l/\left(0.6m_1\right)^2\right)$$

As, $P_1 = P_2$ and $t_1 = t_2$, so

$$W_1/W_2 = \left(\left(-2 \times 10^{-10} V_1^2 + 1 \times 10^{-5} V_1 - 0.03\right)\left(0.6^2\right)\right)/$$

$$2\left(\left(-2 \times 10^{-10} \left(0.5 \times V_1\right)^2 + 1 \times 10^{-5} \left(0.5 \times V_1\right) - 0.03\right)\right)$$

Now, take the value of $V = 200$ mm³

$$W_1/W_2 = \left(\left(-2 \times 10^{-10} \times 200^2 + 1 \times 10^{-5} \times 200 - 0.03\right)\left(0.6^2\right)\right)/$$

$$2\left(\left(-2 \times 10^{-10} \left(0.5 \times 200\right)^2 + 1 \times 10^{-5} \left(0.5 \times 200\right) - 0.03\right)\right)$$

$$W_1/W_2 = \left(\left(-2\times10^{-10}\times200^2 + 1\times10^{-5}\times200 - 0.03\right)\left(0.6^2\right)\right)/$$

$$2\left(\left(-2\times10^{-10}\left(0.5\times200\right)^2 + 1\times10^{-5}\left(0.5\times200\right) - 0.03\right)\right)$$

$$W_1/W_2 = 0.174 \,(\text{approx.})$$

$$W_2 = W_1/0.174$$

$$W_2 = 5.75W_1$$

Or,

$$W_2 = \left(1 + 4.75\right) W_1$$

W_2 is 475% higher wear resistive than W_1.

Q2. What will be the surface finish of the casting, produced by a fused deposition modelling-investment casting process, if the barrel finishing time (t) and media weight (m) are doubled and tripled, respectively? Take the surface roughness model as $\left[6\times10^{-12}(V^2) - 3\times10^{-7}(V) - 0.0082\right].(t.l.\rho^{1/6})/P^{1/3}.m^{2/3}$. Assume required values.

Solution
Now, let us reconsider the equation of surface roughness,

$$R_a = \left[(6\times10^{-12}\left(V^2\right) - 3\times10^{-7}\left(V\right) - 0.0082\right].(t.l.\rho^{1/6})/P^{1/3}.m^{2/3}$$

In case 1: Let the surface roughness of the resulting casting, volume, density, filament strength, finishing time and media weight be R_{a1}, V_1, ρ_1, P_1, t_1 and m_1.

$$R_{a1} = \left[(6\times10^{-12}\left(V_1^2\right) - 3\times10^{-7}\left(V_1\right) - 0.0082\right].t_1\,(\rho_1^{1/6}/P_1^{1/3}).m_1^{2/3}$$

In case 2: Let the surface roughness of the resulting casting, volume, density, filament strength, finishing time and media weight be R_{a2}, V_2, ρ_2, P_2, $2t_1$ and $3m_1$ (as per the given statement).

$$R_{a2} = \left[(6\times10^{-12}\left(V_2^2\right) - 3\times10^{-7}\left(V_2\right) - 0.0082\right].2t_1.(\rho_2^{1/6}/P_2^{1/3}).\left(3m_1\right)^{2/3}$$

Now, dividing Equation 2.1 by 2.2, as under

$$R_{a1}/R_{a2} = \left[(6\times10^{-12}(V_1^2)-3\times10^{-7}(V_1)-0.0082\right].t_1.(\rho_1^{1/6}/P_1^{1/3}).m_1^{2/3} /$$

$$\left[(6\times10^{-12}(V_2^2)-3\times10^{-7}(V_2)-0.0082\right].2t_1.(\rho_2^{1/6}/P_2^{1/3}).(3m_1)^{2/3}$$

As, $P_1 = P_2$, $\rho_1 = \rho_2$ and $t_1 = t_2$

$$R_{a1}/R_{a2} = \left[(6\times10^{-12}(V_1^2)-3\times10^{-7}(V_1)-0.0082\right]/$$

$$\left[(6\times10^{-12}(V_2^2)-3\times10^{-7}(V_2)-0.0082\right]\times2\times(3)^{2/3}$$

$$R_{a1}/R_{a2} = \left[(6\times10^{-12}(V_1^2)-3\times10^{-7}(V_1)-0.0082\right]/$$

$$\left[(6\times10^{-12}(V_2^2)-3\times10^{-7}(V_2)-0.0082\right]\times2\times(3)^{2/3}$$

Now, take the value of $V = 300$ mm³

$$R_{a1}/R_{a2} = \left[(6\times10^{-12}\times(300)^2-3\times10^{-7}(300)-0.0082\right]/$$

$$2\times2.08\left[(6\times10^{-12}\times(300)^2-3\times10^{-7}(300)-0.0082\right]$$

$$R_{a1}/R_{a2} = 0.24\,(\text{approx.})$$

$$R_{a2} = R_{a1}/0.24$$

$$R_{a2} = 4.16R_{a1}$$

Or,

$$R_{a2} = (1+3.16)R_{a1}$$

R_{a2} is 316% higher wear resistive than R_{a1}.

Q3. It is required that the hardness, H, (following [(30 + 13(ρ²) − 40 + 08(ρ) + 2190.7]P².L.t².V/m⁴ model) of the functionally graded material, produced by a fused deposition modelling-investment casting process, should be 1200 N/mm². What should be the filament strength (P) if the values of other parameters are under the following?

- Density (ρ) = 9.163 × 10⁻⁶ N/mm³
- Thickness of the mould (l) = 11.5 mm
- Barrel finishing time (t) = 1200 sec

- Pattern size (V) = 1500 mm³
- Media weight (m) = 98N

Solution

By filling the given values in the following equation:

$$H = \left[(30 + 13(\rho^2) - 40 + 08(\rho) + 2190.7\right] P^2.L.t^2.V/m^4$$

Putting the given values in the earlier equation – that is, $\rho = 9.163 \times 10^{-6}$ N/mm³, L = 11.5 mm, t = 1200 sec, V = 1500 mm³ and m = 98N.

$$1200 = \left[(30 + 13\,(9.163 \times 10^{-6})^2 - 40 + 08\,(9.163 \times 10^{-6}) + 2190.7\right]$$

$$\times P^2 \times 11.5 \times 1200^2 \times 1500/(98)^4$$

$$1200 = 2180.7 P^2 \times 11.5 \times 1200^2 \times 1500/(98)^4$$

$$1200 \times (98)^4/(11.5 \times 2180.7 \times 1200 \times 1200 \times 1500) = P^2$$

$$P = 0.0451 \text{ N/mm}^2$$

or

$$= 0.5 \text{ kgf (approx.)}$$

Q4. Find out the dimensional deviation of the casting produced via a fused deposition modelling assisted investment casting process, under the following set of inputs if the dimensional deviation follows $[-4 \times 10^{-6}(m^2) + 0.0012(m) - 0.048]\rho.V/P^2.l.t^4$ equation.
- Filament strength (P) = 21.65 N/mm²
- Media weight (m) = 196 N
- Pattern size (V) = 14,000 mm³
- Density (ρ) = 7.63 × 10⁻⁶ N/mm³
- Thickness of the mould (l) = 15 mm
- Barrel finishing time (t) = 3600 sec

Solution

Put the given values in the following equation:

$$\Delta d = \left[(-4 \times 10^{-6}(m^2) + 0.0012(m) - 0.048\right]\rho.V/P^2.l.t^4$$

It is given in the statement that; $\rho = 7.63 \times 10^{-6} \text{N/mm}^3$, $l = 15$ mm, $t = 3600$ sec, $V = 14{,}000 \text{ mm}^3$, $P = 21.65 \text{ N/mm}^2$ and $m = 196$ N.

$$\Delta d = \frac{\left[(-4\times10^{-6}(196)^2 + 0.0012(196) - 0.048\right] \times 7.63\times10^{-6}}{1200^2 \times 1500 / (21.65^2 \times 15 \times 3600^4)}$$

$$\Delta d = \frac{\left[(-4\times10^{-6}(196)^2 + 0.0012(196) - 0.048\right] \times 7.63\times10^{-6}}{1200^2 \times 1500 / (21.65^2 \times 15 \times 3600^4)}$$

$$\Delta d = \left[(-4\times10^{-6}(196)^2 + 0.1872\right] \times 5.2986111\times10^{-12} \times 0.00012827771$$

$$= \left[(-4\times10^{-6}(196)^2 + 0.1872\right] \times 6.796937\times10^{-16}$$

$$\Delta d = 2.27\times10^{-16} \text{ mm}$$

Numerical questions (unsolved)

Q1. What will be the effect of wear resistance of the fused deposition modelling assisted investment casting process part if the volume of the pattern (V) is doubled and the thickness of the mould wall (l) is reduced by 50%? Consider the following wear model:

$$W = (-2\times10^{-10}V_1^2 + 1\times10^{-5}V_1 - 0.03)\,\rho_1.P_1.t_1^2.l/m_1^2$$

Q2. Find out the surface roughness of the functionally graded materials prepared with fused deposition modelling patterns in an investment casting process. The following two sets were used:
- Filament strength (P) = 25 N/mm²
- Media weight (m) = 200 N
- Pattern size (V) = 10,000 mm³
- Density (ρ) = 12.63 × 10⁻⁶ N/mm³
- Thickness of the mould (l) = 14 mm
- Barrel finishing time (t) = 2500 sec

Roughness follows equation: $([6 \times 10^{-12}\,(V_1^2) - 3 \times 10^{-7}(V_1) - 0.0082].\,t_1.\,(\rho_1^{1/6}/P_1^{1/3}).m_1^{2/3})$

Q3. Find out the ratio of Vickers hardness (H1/H2) of the functionally graded material produced by a fused deposition modelling-investment casting process under the two different sets of conditions as given in the following table:

Set 1	Set 2
Filament strength (P) = p	Filament strength (P) = 5p
Media weight (m) = 1	Media weight (m) = 0.5
Pattern size (V) = 10X	Pattern size (V) = 20X
Density (ρ) = A	Density (ρ) = 3A
Thickness of the mould (l) = 5 mm	Thickness of the mould (l) = 15 mm
Barrel finishing time (t) = T	Barrel finishing time (t) = 2T

Use hardness trend as given as follows:

$$H = \left[(30 + 13(\rho^2) - 40 + 08(\rho) + 2190.7 \right] P^2.L.t^2.V/m^4$$

Q4. Mathematically calculate the dimensional deviation of the casting (produced via fused deposition modelling assisted investment casting) for given conditions (assume the required values):
Dimensional deviation follows trend:

$$\Delta d = \left[(-4 \times 10^{-6} (m^2) + 0.0012(m) - 0.048 \right] \rho.V/P^2.l.t^4$$

- Thickness of the mould (l) = 12 mm
- Barrel finishing time (t) = 1400 sec
- Pattern size (V) = 2500 mm³
- Media weight (m) = 100 N
- Filament strength (P) = 25 N/mm²

Q5. Find out the ratio (using model as given earlier in unsolved question 2) of roughness (R1/R2) of the functionally graded material under the following two variable sets of parameters:

Condition 1	Condition 2
Filament strength (P) = 0.2p	Filament strength (P) = p
Media weight (m) = 2	Media weight (m) = 0.8
Pattern size (V) = X	Pattern size (V) = 3X
Density (ρ) = 2A	Density (ρ) = A
Thickness of the mould (m) = 1	Thickness of the mould (m) = 2.5
Barrel finishing time (t) = 0.1T	Barrel finishing time (t) = T

References

1. Miracle, D.B. and Donaldson, S.L. 2001. Introduction to composites. *Materials Park, OH: ASM International*, 2001:3–17.
2. Miyajima, T. and Iwai, Y. 2003. Effects of reinforcements on sliding wear behavior of aluminum matrix composites. *Wear*, 255(1):606–616.
3. Prasad, S.V. and Asthana, R. 2004. Aluminum metal-matrix composites for automotive applications: Tribological considerations. *Tribology Letters*, 17:445–453.
4. Kumar, G.V., Rao, C.S.P. and Selvaraj, N. 2011. Mechanical and tribological behavior of particulate reinforced aluminum metal matrix composites– A review. *Journal of Minerals and Materials Characterization and Engineering*, 10:59.
5. Vencl, A., Bobic, I. and Stojanovic, B. 2014. Tribological properties of A356 Al-Si alloy composites under dry sliding conditions. *Industrial Lubrication and Tribology*, 66:66–74.
6. Yılmaz, O. and Buytoz, S. 2001. Abrasive wear of Al_2O_3-reinforced aluminium-based MMCs. *Composites Science and Technology*, 61:2381–2392.
7. Kathiresan, M. and Sornakumar, T. 2010. Friction and wear studies of die cast aluminum alloy-aluminum oxide-reinforced composites. *Industrial Lubrication and Tribology*, 62:361–371.
8. Bermudez, M.D., Martınez-Nicolás, G., Carrion, F.J., Martínez-Mateo, I., Rodríguez, J.A. and Herrera, E.J. 2001. Dry and lubricated wear resistance of mechanically-alloyed aluminium-base sintered composites. *Wear*, 248:178–186.
9. Do Woo, K. and Lee, H.B. 2007. Fabrication of Al alloy matrix composite reinforced with subsive-sized Al_2O_3 particles by the in situ displacement reaction using high-energy ball-milled powder. *Materials Science and Engineering: A*, 449:829–832.
10. Zebarjad, S.M. and Sajjadi, S.A. 2007. Dependency of physical and mechanical properties of mechanical alloyed $Al-Al_2O_3$ composite on milling time. *Materials & Design*, 28:2113–2120.
11. Hassan, S.F. and Gupta, M. 2008. Effect of submicron size Al_2O_3 particulates on microstructural and tensile properties of elemental Mg. *Journal of Alloys and Compounds*, 457:244–250.
12. Gottron, J., Harries, K.A. and Xu, Q. 2014. Creep behaviour of bamboo. *Construction and Building Materials*, 66:79–88.
13. Shen, M. and Bever, M.B. 1972. Gradients in polymeric materials. *Journal of Materials Science*, 7:741–746.
14. Bever, M.B. and Duwez, P.E. 1979. On gradient composites. In: *ARPA Materials Summer Conference*, U.S. Department of Energy's Advanced Research Projects Agency–Energy, USA, pp. 117–140.
15. Kieback, B., Neubrand, A. and Riedel, H. 2003. Processing techniques for functionally graded materials. *Materials Science and Engineering: A*, 362:81–106.
16. Koizumi, M.F.G.M. 1997. FGM activities in Japan. *Composites Part B: Engineering*, 28:1–4.
17. Suresh, S. and Mortensen, A. 1997. Functionally graded metals and metal-ceramic composites: Part 2 Thermomechanical behaviour. *International Materials Reviews*, 42:85–116.

18. Holmström, J., Partanen, J., Tuomi, J. and Walter, M. 2010. Rapid manufacturing in the spare parts supply chain: Alternative approaches to capacity deployment. *Journal of Manufacturing Technology Management*, 21:687–697.
19. Kruth, J.P., Leu, M.C. and Nakagawa, T. 1998. Progress in additive manufacturing and rapid prototyping. *CIRP Annals-Manufacturing Technology*, 47:525–540.
20. Gibson, I., Rosen, D.W. and Stucker, B. 2010. *Additive Manufacturing Technologies* (Vol. 238). New York: Springer.
21. Choy, S.Y., Sun, C.N., Leong, K.F., Tan, K.E. and Wei, J. 2016. Functionally graded material by additive manufacturing. *Proceedings of the 2nd International Conference on Progress in Additive Manufacturing* (ProAM 2016), Singapore, pp. 206–211.
22. Kalita, S.J., Bose, S., Hosick, H.L. and Bandyopadhyay, A. 2003. Development of controlled porosity polymer-ceramic composite scaffolds via fused deposition modeling. *Materials Science and Engineering: C*, 23:611–620.
23. Chung, H. and Das, S. 2008. Functionally graded Nylon-11/silica nanocomposites produced by selective laser sintering. *Materials Science and Engineering: A*, 487:251–257.
24. Sudarmadji, N., Tan, J.Y., Leong, K.F., Chua, C.K. and Loh, Y.T. 2011. Investigation of the mechanical properties and porosity relationships in selective laser-sintered polyhedral for functionally graded scaffolds. *Acta Biomaterialia*, 7:530–537.
25. Mumtaz, K.A. and Hopkinson, N. 2007. Laser melting functionally graded composition of Waspaloy® and Zirconia powders. *Journal of Materials Science*, 42:7647–7656.
26. Hazlehurst, K.B., Wang, C.J. and Stanford, M. 2014. An investigation into the flexural characteristics of functionally graded cobalt chrome femoral stems manufactured using selective laser melting. *Materials & Design*, 60:177–183.
27. Zhang, Y., Wei, Z., Shi, L. and Xi, M. 2008. Characterization of laser powder deposited Ti–TiC composites and functional gradient materials. *Journal of Materials Processing Technology*, 206:438–444.
28. Durejko, T., Ziętala, M., Polkowski, W. and Czujko, T. 2014. Thin wall tubes with $Fe_3Al/SS_{316}L$ graded structure obtained by using laser engineered net shaping technology. *Materials & Design*, 63:766–774.
29. Yang, S., Xue, X., Lou, S. and Lu, F. 2007. Investigation on gradient material fabrication with electron beam melting based on scanning track control. *China Welding*, 16:19–22.
30. Parthasarathy, J., Starly, B. and Raman, S. 2011. A design for the additive manufacture of functionally graded porous structures with tailored mechanical properties for biomedical applications. *Journal of Manufacturing Processes*, 13:160–170.
31. Singh, R. and Singh, S. 2017. Investigating the friction coefficient in functionally graded rapid prototyping of $Al–Al_2O_3$ composite prepared by fused deposition modelling. *Assembly Automation*, 37:154–161.
32. Singh, S. and Singh, R. 2016. Effect of process parameters on micro hardness of $Al–Al_2O_3$ composite prepared using an alternative reinforced pattern in fused deposition modelling assisted investment casting. *Robotics and Computer-Integrated Manufacturing*, 37:162–169.

33. Singh, R., Singh, J. and Singh, S. 2016. Investigation for dimensional accuracy of AMC prepared by FDM assisted investment casting using nylon-6 waste based reinforced filament. *Measurement*, 78:253–259.

34. Singh, S. and Singh, R. 2017. Some investigations on surface roughness of aluminium metal composite primed by fused deposition modeling-assisted investment casting using reinforced filament. *Journal of the Brazilian Society of Mechanical Sciences and Engineering*, 39:471–479.

35. Singh, S. and Singh, R. 2015. Wear modelling of Al-Al$_2$O$_3$ functionally graded material prepared by FDM assisted investment castings using dimensionless analysis. *Journal of Manufacturing Processes*, 20:507–514.

36. Lee, C.W., Chua, C.K., Cheah, C.M., Tan, L.H. and Feng, C. 2004. Rapid investment casting: Direct and indirect approaches via fused deposition modelling. *The International Journal of Advanced Manufacturing Technology*, 23:93–101.

37. Grimm, T. 2003. Fused deposition modelling: A technology evaluation. *Time-Compression Technologies*, 11:1–6.

38. Singh, R. and Singh, S. 2016. Development of functionally graded material through the fused deposition modeling-assisted investment casting process: A case study. In: S. Hashmi (Ed.), *Reference Module in Materials Science and Materials Engineering*. Oxford, UK: Elsevier.

39. *Standard Test Method for Melt Flow Rates of Thermoplastics by Extrusion Plastometer*, ASTM D1238–13.

40. Wang, J. 2007. Predictive depth of jet penetration models for abrasive water-jet cutting of alumina ceramics. *International Journal of Mechanical Sciences*, 49:306–316.

41. Anders, D., Münker, T., Artel, J. and Weinberg, K. 2012. A dimensional analysis of front-end bending in plate rolling applications. *Journal of Materials Processing Technology*, 212:1387–1398.

42. Singh, R. and Khamba, J.S. 2008. Mathematical modelling of surface roughness in ultrasonic machining of titanium using Buckingham-Π approach: A review. *International Journal of Abrasive Technology*, 2:3–24.

chapter two

Metal additive manufacturing using lasers

C. P. Paul, A. N. Jinoop and K. S. Bindra

Contents

2.1   Introduction

Manufacturing is the process of converting raw materials, components, or parts into finished goods as per predefined specifications using labours and tools through various processing techniques. When metals are involved as one of predominant raw materials in this process, it is metal-based manufacturing. This manufacturing sector experienced numerous changes from the pre-industrial revolution era, and it transformed to modern-day manufacturing. Prior to the industrial revolution, manufacturing was limited to artistic form and was often done in homes using hand tools or animal-driven machines. This was the era of *customized product*. Slowly, it witnessed the change towards standardization of parts from *match to fit*. In the first industrial revolution when the conversion of thermal energy into mechanical energy (i.e. steam engine) was developed, manufacturing marked a shift to powered, special-purpose machinery, factories and mass production. The second industrial revolution brought the generation and application of electrical energy into manufacturing. Many of the changes that occurred during this period had to do with new products simply replacing old ones. The third revolution brought forward the computer integrated manufacturing system in the 1970s when the idea of digital manufacturing came into existence. Various functional activities of the factory such as design, planning and inventory were associated with the computer. It also allowed various processes to communicate with each other, exchange information and initiate actions. It reduced the lead time, total cost, inventory and improved the quality and performance. From the first industrial revolution to the third revolution, *mass production* (means manufacturing of similar components in bulk)' was the centreline. Subsequently, it started moving towards a customer-centric approach and brought the idea of *customized production* (means individual components for customer-based requirements) for customer satisfaction bringing a competitive advantage over other products. The industries strived hard to achieve a fine balance between the customer needs and offered product. The narrowing gap between the customer needs and offered product improved the customer satisfaction significantly. Hence, mass customization is recognized by certain companies to target customized products in larger scale. It developed new opportunities by combining two opposing terms in manufacturing called 'mass production' and 'customized production'. It could only be realized due to introduction of cyber-physical systems (integrations of computation, networking and physical processes). This is considered to be the beginning of the fourth industrial revolution with breakthroughs in emerging technologies including robotics, artificial intelligence, nanotechnology, quantum computing, biotechnology, internet, additive manufacturing and autonomous vehicles. Hence, additive manufacturing is one of the important technologies catering to the needs of mass customization through the facilities and expertise available through cyber-physical systems.

Additive manufacturing (AM) is the process of joining materials to make objects from 3D model data, usually layer upon layer, as opposed to subtractive manufacturing methodologies [1]. As we know, AM basically involves three stages such as preprocessing, processing and post-processing. AM provides flexibility for mass customization in all the three stages. In the preprocessing stage, a 3D CAD model is developed and converted to machine-specific format for slicing and numerical control code generation. The 3D CAD model provides the first stage of mass customization. The customer can be directly involved in the design of the component, and the customer has the final say in how the product looks. During the processing stage, many components can be fabricated in a single step with customized design or geometry. The number of components that are built in a single step is governed by the build volume of the AM system. It also adds value to the component by creating something that is a purely unique salient feature with a range of materials. At the end of the processing stage, every product becomes an independent entity, and post-processing can be performed as per the customers' requirements. Due to mass customization, it raised the customer satisfaction to customer delight. AM was initially seen as a process suitable only for concept modelling and rapid prototyping, but it rewrote the possibilities over the years and expanded to build near net-shaped metallic components almost ready to use. The segment of new generation AM processes that are capable of delivering metallic components using standard additive manufacturing topology are called 'Metal Additive Manufacturing (MAM)', more commonly known as Metal 3-D printing. As metals are primarily used as raw material for all engineering applications, MAM added an important dimension to manufacturing. It provides the unique advantages of light weighting, multiple part consolidation, quick design iterations etc. MAM systems can be classified by the energy source or the way the material is being joined such as laser, electron beam, plasma, etc. MAM is progressing to be the next big thing in the industry and thereby transforming the manufacturing sector. It possesses the unique advantage of fabrication of complex-shaped metallic components using various processes and metallic materials, which will be discussed later in this chapter.

2.2 Lasers in metal additive manufacturing

Lasers, being one of the most remarkable inventions of the last century, demonstrated with many exciting applications. The basic concepts of laser were first given by an American scientist, Charles Hard Townes and two Soviet scientists, Alexander Mikhailovich Prokhorov and Nikolai Gennediyevich Basov, who shared the coveted Nobel Prize (1964). However, Theodore Maiman of the Hughes Research Laboratories, California, was the first scientist who experimentally demonstrated laser by flashing light through a ruby crystal in 1960. A laser is a device that emits electromagnetic radiation through a process of optical amplification based on the stimulated emission of photons [2]. The difference between laser and light can be understood with an analogy. When a large number of pebbles are thrown into a pool at the same time, each pebble generates a wave of its own. Since the pebbles are thrown at random, the waves generated by all the pebbles cancel each other and as a result they travel a very short distance. On the other hand, if the pebbles are thrown into a pool one by one at the same place and also at constant intervals of time, the waves thus generated strengthen each other and travel long distances. In this case, the waves are said to travel coherently. In laser, the light waves are exactly in step with each other and thus have a fixed phase relationship [3]. These relationships make lasers suitable for materials processing under different specialized conditions as lasers provide coherence, both spatial and temporal; high intensities and fluence capabilities; short to ultra-short pulse durations; and there is no need of a vacuum for its propagation and feasibility of transportation through flexible optical fibres. Due to these unique properties, lasers have established themselves in materials processing not only from a utilitarian point of view but also as a vibrant research tool to seek solutions for future requirements. The use of lasers in the manufacturing industry has several advantages over conventional methods. Some of them are listed as follows:

1. Laser-assisted manufacturing techniques are a noncontact process and are well suited for processing advanced engineering materials such as brittle materials, soft and thin materials, electric and non-electric conductors, etc.
2. It is a thermal process, and materials with favourable thermal properties can be successfully processed regardless of their mechanical properties.
3. It is a flexible process [3].

The right choice of laser and proper selection of processing parameters are mandatory to determine the applicability and capability of a laser in any application. Though a large variety of lasers employing different kinds of active mediums covering a wide range of wavelengths and powers has been developed, only a few are being used for materials processing. CO_2, Nd:YAG, fibre lasers and diode lasers are the most popular systems.

The lasers were first to provide the biggest leap to AM technologies to extend it to MAM, placing it into a group known as 'disruptive technologies'.

MAM involving lasers is known as Laser Additive Manufacturing (LAM). Shapes and designs that could not previously even be conceived of are now entirely viable. Parts with conformal cooling channels are one of such examples. It is not just the shape and design that became possible, but LAM could also keep the advantages in terms of the structural properties. There is no weakening of the structure and no need to join several parts together, creating failure-prone joint areas. Instead, the LAM can build a complete shape in its entirety by carefully layering and bonding materials together to create the desired outcome in a single step. LAM derived a conceptual revolution due to the following five exclusive freedoms:

1. *Material design freedom*: The concept of layer-by-layer manufacturing provided the freedom of introducing different material at various locations within a component. Materials having metallurgical incompatibility can be brought together by inserting a friendly buffer layer between the two. The functional grading of materials provided the graded material transition across the material interface at the dissimilar material joint and thereby increased the interface strength. These are the some of the examples of material design freedom provided by LAM.

2. *Shape design freedom*: The conceptual revolution of LAM provided new possibilities in component design and manufacturing by providing complex geometries with undercuts and internal spaces, lighter and honeycomb with reduced weight and material usage and single pieces without joining of individual components. The availability of advanced computational tools provided the insight of stress and flow fields that in turn changed the component design, retaining material only in locations where it affects the functionality of the part. It increases the complexity but reduces material and weight. Less material is added to produce the lightweight part and thus, manufacturing time and costs can be saved. In contrast to conventional manufacturing, increased complexity in LAM does not lead to higher cost. It is particularly attractive for high-value materials.

3. *Logistics freedom*: LAM reduced the gap between manufacturer and consumer by directly taking input for mass customisation. The availability of cyber technologies enabled the design, manufacturing and testing of the components at far and remote locations. The availability of global expertise locally replaced many parts with single parts of equivalent geometry. It redefined the supply-chain management by reducing the number of agencies involved. LAM technology is fully compatible to INDUSTRY 4.0.

4. *Quality freedom*: Quality concept is altered due to simultaneous building of geometry and material. Rapid deposition ensured the consistency in material and geometry quality. It changed the material and geometry validation to process validation. Due to in situ inspection, quality became an integral part of process transferring into part.

5. *Post-processing freedom*: Building strategy can be developed as per post- processing requirements. Dimensionally critical parts can be derived using machining during the design stage itself.

The aforementioned exclusive freedoms made LAM ready to undertake newer 'feature-based design and manufacturing', where the components are fabricated as per the functional requirements by providing optimum solution in terms of material, shape and processing.

2.3 Global scenario of laser additive manufacturing

The global market of 3D printing is expected to grow by 23.2% for the period of 2016–2021 as per research and markets [4]. A global survey published by Metal Powders includes different types of 3D printing techniques and addresses the trends in the technology and the global market [4]. EOS plans to introduce EOSTATE melt pool for the process monitoring of LAM processes, which will be an add-on for new models of EOS M 290 [5]. This will help in online process control and quality inspection. EOS plans to use this tool for aerospace components where tolerances and high performance are necessary. This tool helps to detect the potential defects at an early stage [5]. Siemens is planning to extend its focus on high-temperature super alloys after acquiring the major stake in Materials Solutions Ltd. Siemens has successfully performed serial production for rapid manufacturing of fuel mixers and repairs of turbine blade tips [6]. Siemens fabricated 3D printed turbine blades that passed 13,000 rpmat 1250°C. A polycrystalline nickel superalloy powder that can withstand elevated temperature and pressure was used for fabrication [7]. In addition to this, Siemens and the Fraunhofer Institute for Laser Technology developed a faster process based on a Laser Melting (LM) process. The long lead time in the fabrication of turbine blades and vanes gave the motivation for this approach. Thus, scientists developed a process that speeds up the manufacturing of turbine vanes slated for hot gas regions in the engine. General Electric (GE) plans to invest $1.4 billion in 3D printing and plans to acquire two MAM suppliers, Arcam AB and SLM Solutions [8]. A modular approach for additive manufacturing was developed by Concept Laser. The company developed a process station with a build volume of 400 × 400 × 400 mm with 1–4 laser systems. The power of laser sources varies from 400 to 1000 W with adjustable spot size from 50 to 500 microns. It has a two-axis coater system with a machine data and operating data logging system [9]. Concept Laser GmbH has fabricated a frame for lightweight sports cars called Light Cocoon. It features bodywork nodes made by AM using a laser cusing process with a 1 kW laser [10]. In addition to this, Concept Laser fabricated a titanium replacement beak for a wild macaw. The wild macaw lost its beak because of poor conditions. This prosthetic helped the bird, as macaws are unable to eat solid food without a beak [11]. Biomedical applications of AM were extended by Metalysis, UK, by developing both uniform and randomised tantalum lattice structures with structural stiffness similar to that of bone [12]. Researchers in Singapore have developed a titanium-tantalum alloy powder that can be used to make biocompatible bone replacement through Laser Melting. While successful titanium-aluminium bio prototypes were developed before, the concern over the long-term effects of aluminium on the human body motivated the researchers to go for other materials. Titanium-tantalum alloy is considered as it is biocompatible and has mechanically superior properties compared to titanium [13]. Ferrari is planning to incorporate additive manufacturing to remodel the ignition chamber of the engine. It is believed that additive manufacturing will allow Ferrari to increase the strength to weight ratio of the piston heads. Ferrari may use complex internal geometries such as honeycomb patterns, which results in material reduction while ensuring appropriate strength in high-impact areas [7]. SLM Solutions Group plans a joint venture with PKM Future Holding GmbH to invest in the development and distribution of aluminium alloys for MAM systems [14]. Altair and Renishaw have designed and manufactured a mountain bike frame using a MAM technique. It used titanium lugs, carbon fibre components and tubings with double-lap bonding [15]. The world's first 3D printed motorcycle was launched by airbus. It is 30% lighter than conventionally manufactured e-motor cycles. A special algorithm was employed to obtain required strength to weight ratio. Bionic design was realised for the motorcycle and thus, it eliminated the limitations of conventional techniques. Scalmalloy is an aluminium alloy as strong as titanium with high corrosion resistance, making it suitable for various industrial applications [16]. Researchers have developed a new way to fabricate metallic parts using ink with a variety of metals, metal mixtures, alloys and metal oxides and compounds. Liquid ink with metals or mixed metal powders, solvents and binder was employed for printing components, bypassing the powder bed and energy beam approach [17].

To meet the requirements of experts in the area of additive manufacturing, academia is also preparing a dedicated subject at the undergraduate level, and specialization at the post-graduate level is being introduced. The National Institute of Technology, Warangal, India, started a Master of Technology program in Additive Manufacturing in the year 2014 [18]. Along with this, Anglia Ruskin University in the UK has started the UK's only dedicated master's degree in additive manufacturing [19]. Arizona State University in the US is planning to start a large additive manufacturing facility in association with Concept Laser, Honeywell Aerospace and Phoenix Analysis & Design Technologies [20]. IESE in Barcelona, will work with EOS to deliver an executive education program on 'Industry 4.0: The Future of Manufacturing' [21]. Johns Hopkins opened an additive manufacturing centre for excellence in the US to guide the major advances in the field of AM. The centre aims to work on design, fabrication, materials and characterization of AM [22].

2.4   Classification of laser additive manufacturing process

A LAM system consists of the following three primary subsystems – a high-power laser system, a material feeding system and a CNC workstation. The CO_2 laser, Nd:YAG laser, Diode laser and Fibre laser are the industrial workhorses. The material feeding system includes the feeding of wires and powder. In the CNC workstation, three- and five-axis workstations as well as galvano-scanners are being used in the LAM system. Based on the earlier, the classification of the LAM system is presented in Figure 2.1.

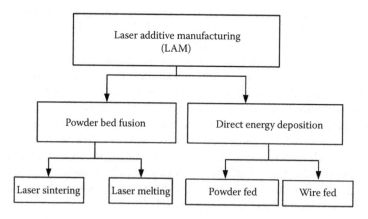

Figure 2.1 Classification of LAM processes.

2.4.1 Powder bed fusion

In a Powder Bed Fusion (PBF) system, a thin layer of metal powder is laid over the bed and a focused laser beam is used to selectively melt or sinter the layer of a powder bed. A number of layers, one over the other, are laid to shape the component as per the design. In the process, the adherence of the current layer with the previous layer is ensured by a seamless binding mechanism. The most commonly used PBF systems are laser sintering and laser melting.

1. *Laser Sintering (LS)* was developed by Carl Deckard for his master's thesis at the University of Texas and patented in 1989 [23]. The first reported metal AM part was made from metal alloy powders in an LS process in 1990 by Manriquez-Frayre and Bourell. A computer-controlled, low-power, Q-switched, TEM_{00}Nd:YAG laser with nominal power capacity of 100 W and beam size of 0.5 mm was used to selectively sinter metal powders. The material used included 99.0% copper, 99.5% tin and 70 Pb-30 Sn [24].

   LS fabricates three-dimensional components using laser energy to powder beds. Like other AM techniques, it also requires a CAD model of the part geometry for initiation, from which it develops a 2-D stack of layers, which can represent the component. The powder is spread over the surface of a build cylinder and the layer of the components is then developed by selectively scanning over the required cross-sectional area using laser and sinter and bond particles in a thin lamina. Excess powder in each layer acts as support for the part during the build. The entire fabrication chamber is sealed and maintained at a temperature just below the melting point of the powder. Thus, heat from the laser need only to elevate the temperature slightly to cause sintering, greatly speeding the process. Subsequently, the build plate moves down by one-layer thickness to accommodate the layer of powder, and another layer is created and bonded to the already created layer. The process is repeated until geometry described by the original solid model is developed [23,25]. Free poured (loose) or slightly compacted powders with particle size in the range of several microns to several-hundred microns are typically used for LS [26]. Polymer coated metal powders are also used in LS. When the laser beam heats the powder bed, the melt-front moves through the low conductive powder layer surface to the bulk. When the melt-front reaches the substrate, high thermal conductivity of the substrate/previous deposited layer improves the heat conduction sharply. If the laser energy is sufficient to melt a thin layer of the substrate/previous deposited layer, the melted powder is fused and deposited, forming a metallurgical bond. If excess energy is used, it leads to excessive remelting and thermal distortion, while insufficient energy causes poor adherence and incomplete melting of the powder. For any particular deposited height (h) of a specific material, a specific laser energy (E) can be calculated by

$$E = \frac{P}{V}$$

(2.1)

where:
    P is the laser power
    V is the scan velocity

The value of specific laser energy (E) has to be within a certain range for the controlled remelting of the substrate and to achieve good powder consolidation for a specific material [27].

2. *Laser Melting (LM)*: Metal powder-bed processes with full melting of powders have been named as Selective Laser Melting (SLM), Direct Metal Laser Sintering (DMLS), etc., depending on the vendor. SLM is the most commonly used name for PBF processes. EOS GmbH, SLM Solutions, Concept Laser, Renishaw and 3D Systems are the leading manufacturers of PBF systems with full melting capability. Figure 2.2 represents the schematic of the Laser Melting system.

    The availability of high-power compact solid-state lasers, advanced electronics and materials paved the way to the rapid

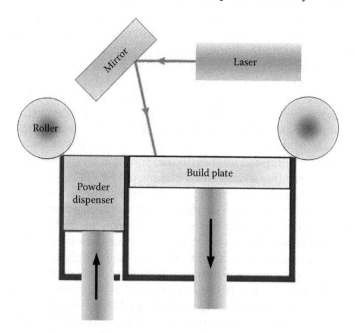

Figure 2.2 Schematic of laser melting process.

development of LM technology since the year 2000. The procedures and apparatus of LM remains the same as that of LS. The only difference is that LM is based on complete melting of metallic powder particles. The requirement to produce fully dense components with mechanical properties comparable to those of bulk materials and the desire to evade time consumption in post-processing motivated the development of the LM process. Even though the processing parameters are different for each metal considering laser absorption, surface tension and viscosity of the liquid metal, LM is a candidate for all metals. This is achieved by developing a process window experimentally for each metal to avoid defects such as balling effects, porosity, etc. [28]. The amount of energy transferred to the powder determines the degree of sintering and melting of powder particles to produce solid parts. The energy transferred to the feedstock normally is primarily influenced by the laser power, the scan speed and scan spacing. Gu and Shen [29] demonstrated different scanning strategies and reported that an optimal amount of laser power and scanning speed are required for the complete melting of powder particles.

2.4.2 Direct energy deposition

Although the processing strategy of direct energy deposition (DED) follows the general additive manufacturing principle, the mechanism of feedstock supply is different from the PBF process. PBF systems employ pre-spreading of feedstock while the DED system follows a feedstock feeding mechanism. DED includes all methods where feedstock is deposited onto a melt pool generated by a focused beam of energy, such as a laser or electron beam. The initiation of this technology is from welding, where material can be deposited outside a build area by a flowing shield gas [30]. One of the most commercialized DED systems is patented as the LENS process developed at Sandia National Laboratories [31]. LENS can be considered a disruptive technology that may be used for repairs and freeform fabrications. The word disruptive is used since the technology challenges one to think out of the box because of its unique capabilities. It combines excellent material properties with near-net-shape, direct-from-CAD, part building and repair, which makes it unique as compared to other additive processes. DED processes, which feed wire into a molten pool (wire-fed), are essentially extensions of welding technology. Thus, the DED process can be powder fed deposition (PFD) or wire fed deposition (WFD).

1. *Powder Fed Deposition*: Powder feeding is convenient and the most widely used approach for material feeding in DED systems. It allows online variation in feed rate and multi-material feeding. Moreover, laser energy utilization is also greater in dynamic powder blowing,

as the laser beam passes through the powder cloud to the substrate/ previously deposited layer, resulting in the preheating of powder particles through multiple internal reflection and absorption of laser beam. The rate of deposition and properties of the components are influenced by the angle of powder spaying in PFD. Deposition rate and velocity of heat source influence the amount of material deposited in a given pass. In PFD, the build-up of the material must be considered to appropriately define the z-axis layer size or thickness [27]. Figure 2.3 presents the schematic of a typical PFD system. High-power lasers commonly deployed for the PFD applications include diode laser, fibre laser, Nd:YAG laser and CO_2 laser [27].

Powder delivery system consists of a powder feeder that delivers powder into a gas delivery system. The powder is carried to the deposition zone by a carrier gas, while a shielding gas protects the melt-pool from the influence of the atmosphere. Helium and argon are the most commonly used shielding gases while in some cases a mixture of both can be used to combine the benefits of both gases – that is, the higher density of argon and the higher ionisation potential of helium to obtain improved shielding of the melt zone. The high-energy laser beam is delivered via coaxial nozzles along the z-axis in the centre of the nozzle array and focused by a lens in close proximity to the workpiece. Height of the focuses of both laser and powder is manipulated by moving the lens and nozzles in the z-direction. The workpiece/laser head is moved in the x–y direction by a computer-controlled drive system under the beam/

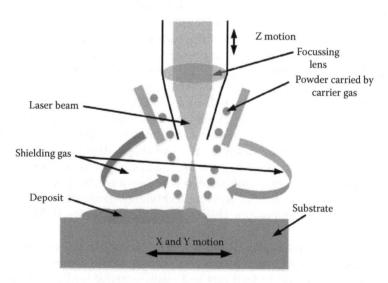

Figure 2.3 Powder feed LAM system.

powder interaction zone to form the desired cross-sectional geometry. Consecutive layers are additively deposited, producing a three-dimensional component.

2. *Wire Fed Deposition*: Wire fed deposition (WFD) is generally used in various industrial applications, such as gas metal arc welding and submerged arc welding. The standard wire feeders used in this technique are customized and used in many LAM systems. In the WFD technique, material is deposited by feeding wire through a nozzle and melting the fed wire with a laser beam. The molten metal forms a metallurgical bond with the substrate/previously deposited material – and by relative motion between the substrate and laser head/wire feeder, a metal track is generated after solidification of molten metal. To protect the oxidation of molten metal, the nozzle is integrated with either a suitable inert gas shielding arrangement or a controlled atmosphere chamber. The advantage of WFD over PFD lies over the fact that the deposition rate is almost 100% with minimal material wastage. In addition to this, less health hazards may occur as compared to PFD. Figure 2.4 presents a schematic of a typical lateral Wire feed LAM System.

WFD has additional process parameters as compared to other PFD system such as the angle between the laser beam and the feed wire, wire diameter and wire tip position. The angle between the laser beam and the feed wire influences the wire feed rate, standoff distance and simultaneously the reflection of the beam, thereby the energy absorbed. Wire diameter should also be properly selected based on the laser diameter for effective melting

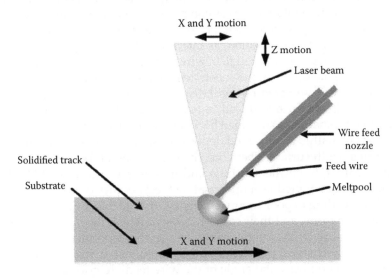

Figure 2.4 Lateral wire feed LAM system.

and deposition. The position of the wire tip relative to the melt pool also influences the melting of the wire and thereby the stability of the process. WFD systems are highly sensitive to process parameters and thus the balancing of processing parameters such as wire tip position in the melt-pool, feed angle, feed direction, laser spot size, laser power (P), wire feed rate (WFR) and traverse speed (V) are important for stable deposit. Front feeding and rear feeding can be deployed for WFD systems [32]. Front feeding involves feeding of wire in the opposite direction of deposition, while in rear feeding wire is fed in the direction of deposition. In front feeding, the surface roughness increases with increased feeding angles, while for rear feeding the surface roughness decreases with increased feeding angles. As mentioned before, the position of feed wire is also significant in achieving a regular and uniform deposit. Good deposit is obtained in front feeding and rear feeding when the wire is positioned in the leading edge and trailing edge, respectively. No porosity is observed in front wire feeding, while some porosity is detected in rear wire feeding [32].

One of the recent developments in WFD is a coaxial wire feeding system. A coaxial wire feeding system uses a metallic wire fed coaxially and laser beam to make three-dimensional components. It essentially consists of a coaxial wire feeding nozzle, high power laser beam, beam splitters and beam reflectors. Figure 2.5 presents a schematic of a coaxial wire feed LAM System. The beam is split into three different beams using beam splitters and reflected by reflectors and focused at a single point, keeping horizontal orientation 120° apart. The metallic wire is melted using the focused laser beam and can be used for making complex 3D components from CAD model data. The major advantage of this wire feeding system is the omnidirectionality it offers as compared to lateral wire feeding.

Wire feeding is preferred for the fabrication of components as it involves continuous deposition, as intermittent starts/stops result in discontinuity in deposited material. In wire feeding, the wire is always in contact with the melt-pool on the substrate [27]. Inaccuracy in the positioning of wire and the wire-feed rate will disturb the shape and size of the melt-pool. This disturbance will lead to non-uniform deposition. Moreover, a definite ratio should be sustained between the beam diameter and the wire diameter (usually >3) for a good deposit substrate metallurgical bond with uniform smooth surface. Therefore, the positioning of the wire in relation to the beam spot on the substrate and its size are critical in WFD. This method is not adopted as a universal method because of poor wire/laser coupling, leading to poor energy efficiency, unavailability of various materials in wire forms and their cost. The way in which the material is delivered to the surface has a significant effect on the deposition rate, material defect and mechanical properties of the component. In a wire fed system, the vertical and horizontal angle of the wire feed are governing the laser energy coupling, surface roughness and melting of the feedstock.

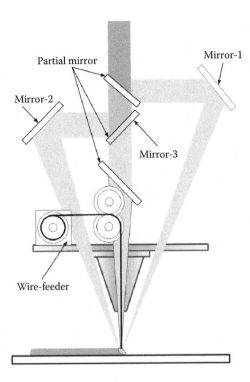

Partial mirror

Mirror-1

Mirror-2

Mirror-3

Wire-feeder

Figure 2.5 Coaxial wire-feeding nozzle.

2.5 Components of laser additive manufacturing system

2.5.1 Laser system

LAM uses a high-power laser system as a heat source for melting a thin layer of substrate, deposited material and fed or preplaced material. CO_2 laser, Nd:YAG, fibre and diode lasers are most widely used for LAM. Among various lasers, shorter wavelength lasers such as Nd:YAG, fibre and diode lasers are most common for MAM due to better absorption, while CO_2 lasers are still common in spite of its larger wavelength due to established procedures and systems [27]. Lasers of about 200 W to 6 kW of various wavelengths are used in commercially available systems. To further extend the capabilities of LAM, more powerful lasers with improved wall plug efficiency and better overall process control are necessary. This may improve the process repeatability, part reproducibility, deposition rate and high-energy use. Table 2.1 presents some major manufacturers of LAM equipment and their lasers.

Table 2.1 Some commercial LAM manufacturers, model and system components

Category	Manufacturer	Model	Laser	Material feeding	Workstation	Build volume
Powder fed deposition	Optomech	LENS 450	400 W IPG Fibre Laser	One integrated feeder (option to add one more)	3-axes standard: XY linear table motion Z Gantry motion	$100 \times 100 \times 100$ mm^3
		LENS MR-7	500 W Fibre Laser	Dual Powder Feeders (Each 2 L)	3-axes standard: XY linear table motion Z gantry motion Additional Axis Optional rotary axis All axes under full CNC control	$300 \times 300 \times 300$ mm^3
		LENS 850-R	1 kW Fibre Laser	Two feeders each hold up to 14 kg of powder	5-axes standard: XYZ linear gantry motion Tilt-Rotate worktable All axes under full CNC control	$900 \times 1500 \times 900$ mm^3
	DM3D technology	DMD 105D	1 kW Fibre coupled laser-Diode/Disc/ Fibre Laser	2 standard 500 cm^3 per hopper	5 axes (x, y, z, B, C)	2D: $800 \times 800 \times 3000$ 3D: $300 \times 300 \times 300$

(Continued)

Table 2.1 (Continued) Some commercial LAM manufacturers, model and system components

Category	Manufacturer	Model	Laser	Material feeding	Workstation	Build volume
Powder bed fusion	EOS	EOS M 100	Yb fibre laser; 200 W	Moving platform	F-theta-lens, high-speed scanner	Ø 100 × 95 mm
		EOS M 290	Yb fibre laser; 400 W	Moving platform	F-theta-lens, high-speed scanner	250 × 250 × 325 mm^3
		EOS M 280	Yb fibre laser, 200 W or 400 W (optional)	Moving platform	F-theta-lens, high-speed scanner	250 × 250 × 325 mm^3
		EOS M 400	Yb fibre laser; 1 kW	Moving platform	F-theta-lens	400 × 400 × 400 mm^3
	SLM solutions	SLM 500	Twin (2x 400 W), Quad (4x 400 W), Twin (2x 700 W), Quad (4x 700 W) IPG fibre laser)	Moving platform	Flying optics	500 × 280 × 365 mm^3
		SLM 280 2.0	Single (1x 400 W), Twin (2x 400 W), Single (1x 700 W), Twin (2x 700 W), Dual (1x 700 W and 1x 1000 W) IPG fibre laser	Moving platform (bi-directional)	Flying optics	280 × 280 × 365 mm^3
		SLM 125	Single (1x 400 W) IPG fibre laser	Moving platform (bi-directional)	Flying optics	125 × 125 × 125 mm^3

(Continued)

Table 2.1 (Continued) Some commercial LAM manufacturers, model and system components

Category	Manufacturer	Model	Laser	Material feeding	Workstation	Build volume
	Concept lasers GmbH	Mlabcusing	Fibre laser 100 W (cw)	Moving platform	Flying optics	$50 \times 50 \times 50$ mm³ (x, y) $70 \times 70 \times 70$ mm³ (x, y) $90 \times 90 \times 90$ mm³ (x, y)
		Mlab cusing 200R	Fibre laser 200 W (cw)	Moving platform	Flying optics	$100 \times 100 \times 100$ mm³ (x, y, z) $70 \times 70 \times 80$ mm³ (x, y, z) $50 \times 50 \times 80$ mm³ (x, y, z)
		M1 cusing	Fibre laser 200 W (cw), optional 400 W (cw)	Moving platform	Flying optics	$250 \times 250 \times 250$ mm³
		M Line Factory	4×400 W or 4×1000 W	Moving platform	Flying optics	$400 \times 400 \times$ up to 425 mm³
		M2 cusing	M2 cusing laser system: Fibre laser 200 W (cw), optional 400 W (cw) M2 cusing Multilaser laser system: Fibre laser 2×200 W (cw)	Moving platform	Flying optics	$250 \times 250 \times 280$ mm³ (x, y, z)
		X LINE 2000R	Fibre laser 2×1 kW (cw)	Moving platform	Flying optics	$800 \times 400 \times 500$ mm³ (x, y, z)

Source: https://www.optomec.com; http://dm3dtech.com; https://www.eos.info; https://www.materialstoday.com; http://www.conceptlaser inc.com [33–37].

2.5.2 Powder feeding system

As mentioned before, LAM can be classified into PBF and PFD. In PBF, pre-spreading of powder is done on the substrate, and the laser selectively melts the powder according to the required geometry. Powder is fed into the deposition zone in the PFD system using powder feeding mechanisms. These powder feeding mechanisms are discussed in this section.

In PBF systems, the powder material is laid on the surface of the substrate with thickness equal to one-layer thickness. The mechanism for powder delivery in the PBF system varies from manufacturer to manufacturer. But commercially the most common material delivery systems are:

1. Moving platform-based powder feeding
2. Hopper-based powder feeding

A moving platform-based powder feeding system uses a powder supply bin filled with sufficient powder to build the part till its maximum height. Figure 2.6 presents the moving platform-based powder feeding. Powder is supplied by the powder supply bin that moves powder upwards in incremental steps. Once the powder stock is raised, the blade or roller spreads it over the build table and pushes any excess powder into the overflow bin. The thickness of the layer depends on the movement of the build platform since the roller height remains constant. The roller moves in a counterclockwise fashion, which helps in fluidizing the powder particles and improves its fluidity.

A hopper-based powder feeding system essentially uses a hopper, which is filled by a feeding system. Figure 2.7 presents the hopper-based

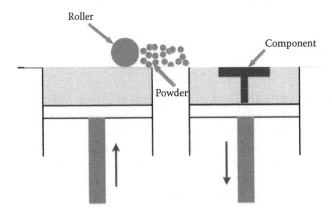

Figure 2.6 Moving platform-based powder feeding.

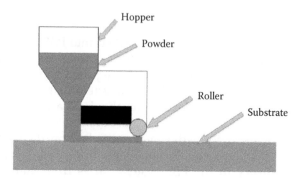

Figure 2.7 Hopper-based powder feeding.

powder feeding system. The powder is delivered from above the processing area in front of a roller or blade, which combines the feeding and
spreading of powder. It is achieved by moving the build platform down-
wards with the hopper remaining stationary. The downward movement
of the platform of the build platform is equivalent to the layer thickness.
The deposited powder is spread and levelled thereafter by roller or blade.
In addition to this, the roller or blade can be integrated with the feeding
system for combined feeding and spreading. Ultrasonic vibration can
be deployed with both the approaches for easy delivery of the powders.
A hopper-based powder feeding system can be a filled hopper and dosimeter hopper. In a filled hopper, the powder is filled and spread over the bed,
while the defined quantity of powder is filled in the dosimeter hopper with
the aid of a dosimeter.

A PBF system also employs the powder levelling mechanism for the
uniform powder spreading on the substrate surface. They are mainly of
two types:

1. Roller type
2. Blade type

The roller type powder levelling mechanism uses a roller to spread the
powder layer on the substrate surface. The roller rotates in a counterclockwise fashion and pushes the powder in the upward direction. The powder
is spread over with minimal shear stress on the previous layer. A blade
type powder leveller is presented in Figure 2.8. It deploys a metal piece
to spread the powder across the bed. Contrary to the roller type, a blade
type powder levelling system induces more shear to the previous layer
with minimal fluidity.

Figure 2.8 Blade-type powder levelling system.

In a PFD system, powder carried by carrier gas is fed online to the deposition zone for melting and deposition in a layer-by-layer fashion. Various powder feeder configurations are used and can be classified based on the working principles as the following:

1. Pneumatic screw-type powder feeder
2. Vibration assisted gravity powder feeder
3. Volumetric controlled powder feeder

A pneumatic screw-type powder feeder is one of the oldest methods. The powder is fed by a rotating screw mounted at the bottom of the hopper, and the controlled amount of the powder is fed into a pneumatic line and transported to the powder nozzle with an inert gas like Argon (Ar), Helium (He), etc. The rotational speed and dimensions of the screw control the powder feed rate. Frequent choking and the need for cleaning are the problems observed in this feeder. In a vibration assisted gravity powder feeder, a combination of pneumatic and vibration forces is used to feed the powder into a pneumatic line. A hopper carrying powder and a vibrating device are the most common parts of this feeder system. A vibrating device can be a rotating wheel or standard ultrasonic vibrator and is fitted at the bottom of the hopper. The powder flows freely through an orifice. The dimensions of the orifice control the powder flow rate. The vibrating device generates the vibration and maintains the free flow of powder under gravity into the pneumatic line. Thus, the powder is delivered to the substrate with the help of an inert gas. This approach has limitations in regulating the powder flow rate in finer steps, and it is widely used for applications involving a constant powder flow rate. In volumetric powder feeder, the powder is filled in the hopper and it falls on the annular slot of a rotating disc. The hopper and the chamber that encloses the

Table 2.2 Comparison of lateral and coaxial nozzle

Parameter	Coaxial	Lateral
Positioning accuracy	Accurate positioning	Difficult to align
Powder utilization efficiency	Independent of traverse speed	Dependent on traverse speed
Part complexity	More	Less
Part quality	Smooth, uniform and consistent	Rough cladding track

Source: Paul, C.P. et al., Laser rapid manufacturing: Technology, applications, modelling and future prospects, in Davim, J.P. (Ed.), *Lasers in Manufacturing*, Wiley, Hoboken, NJ, 1–60, 2012; Toyserkani, E. et al., *Laser Cladding*, CRC Press, New York, 2004 [38,27].

rotating disc are pressurised equally. During the rotation of the wheel the powder is carried to a bore, where a gas flow, created by an imposed pressure difference, pushes it into a connecting tube and powder is delivered to a desired point with the help of a flexible tube. The further details regarding PFD powder feeding systems are published elsewhere [27].

In the PFD systems, there are two basic configurations for powder delivery to the substrate – that is, lateral and coaxial powder feedings. Lateral feeding is very common in laser cladding [27]. Since the powder stream is injected off-axis from the laser beam in lateral feeding, substrate movement determines different local geometry of deposition. Table 2.2 presents the comparison of coaxial and lateral nozzle deposition systems.

The coaxial nozzle consists of three concentric polar matrices of holes. The central passage is used for laser beam and shielding gas; the next outer passages are used to inject the powder with the carrier gas in the converging cone shape and form a coaxial laser-powder stream; the outermost passages are used to supply a converging conical jet of shielding gas.

2.5.3 Beam and job manipulation system

The beam and job manipulation system scans the laser beam for the selective deposition as per the NC code developed from a solid model. The configuration and accuracy of beam and job manipulation system reflects in the dimensional accuracy and reproducibility of the LAM-built components. Different types of beam and job manipulation systems are used for powder fed and powder bed configuration systems. This section describes the configuration of these systems.

2.5.3.1 Powder fed deposition

A CNC workstation is commonly used in a PFD system for job manipulation. Three-axis interpolation (X, Y and Z) is enough for reaching any

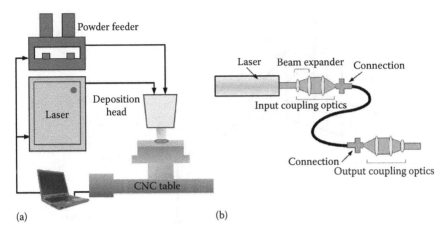

Figure 2.9 (a) Schematic of PFD-based LAM system and (b) beam manipulation in PFD LAM.

point in the area of deposition, but the additional requirement of two more axis (A and C) is to orientate towards a specific direction. Thus, a five-axis configuration is a universal requirement without any redundancy. A laser beam is fed through a beam expander and fibre is used for transmitting the beam from one end to other. The need of a beam expander will be explained later in this section. Figure 2.9 presents the schematic of a PFD-based LAM system and beam manipulation in PFD LAM.

2.5.3.2   Powder bed fusion

In powder bed fusion (PBF) the laser beam is manipulated using moving optics. A beam expander is an optical device, composed of two lenses that produces an output beam with a near-zero divergence. It enlarges the beam diameter by selected amounts. For example, 1.6x, 2x and 2.5 times diameter of the beam are standard expansion ratio. The magnification of the beam expander can be expressed as the diameter of the output beam divided by the diameter of the input beam. Thus, it increases the working distance of a lens and allows the beam to travel a long path without divergence. Figure 2.10 presents beam manipulation in PBF using a galvanoscanner.

Spherical lenses focus the laser beam on a spherical plane in contrast to an ideal flat or plane field. The flat-field scanning lens solves this problem. However, with the absence of distortion, the displacement of the beam depends on the product of the effective focal length (f) and the tangent of the deflection angle θ [$f \times \tan (\theta)$]. In order to maintain uniform exposure on the material being scanned, the constant power image spot must move at a constant velocity. The scanning spot will move along the scan line at a constant velocity if the displacement of the spot is linearly

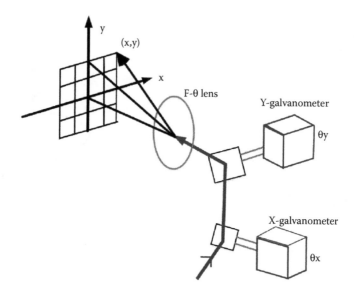

Figure 2.10 Beam manipulation in PBF using galvano scanner.

proportional to the angle θ. The displacement Y of the spot from the optical axis should follow the equation:

$$Y = F\theta$$

F-θ lens is designed with built-in barrel distortion. Thereby, the position of the focused spot can then be made dependent on the product of F and θ, thereby simplifying positioning algorithms. Distortion is the result of the change in the transverse magnification in the system as a function of the distance from the optical axis. When the magnification decreases as the distance from the axis increases, it produces a barrel-shaped image. The use of *f*-theta lenses provides a plane focusing surface and almost constant spot size over the entire XY image plane or scan field.

VarioSCAN is also used in a few systems after the beam expander for laser beam manipulation. It is an optical system with the capability of dynamic variable focal length.

It consists of three major parts:

1. The motor block with its water cooled entrance aperture, its diverging optic and its clamping surface for mounting the VarioSCAN
2. The objective mount
3. The objective with its focusing ring

The beam enters the VarioSCAN from beam expander and is expanded by diverging optic (a negative lens). Next, the beam is focused by the objective,

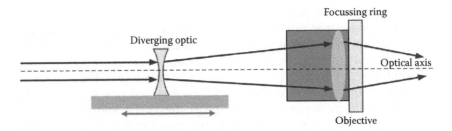

Figure 2.11 Beam manipulation using VarioSCAN.

which consists of a lens system with positive focal length. Variation of focal length is achieved via a motor that moves the mentioned diverging optic along the optical axis under the programmatic control. Working distance can be fine-tuned by turning the focusing ring and thus, moving the objective lens along the optical axis. The beam diameter can be matched to the entrance aperture of the scan head by choosing a beam expander with a suitable expansion factor. Figure 2.11 presents beam manipulation using VarioSCAN.

2.6   Process parameters

The most common process parameters that influence various LAM processes are:

1. *Laser power*: Laser power refers to the rate at which energy is generated by the laser. Laser power affects the laser energy fed to the material, which determines the mechanical properties of the built parts. When the laser power density was lower than the specified range for a given material, the discontinuous tracks were observed with a non-uniform cross-section. On the other hand, excessive remelting of substrate/pre-deposited layer is observed when the laser power density is more than the specified laser power density.
2. *Material feed/flow rate*: It refers to the rate of flow of the feedstock, which finds application in DED systems. In the case of single-track deposition, for a given laser power density and interaction time, the deposition rate increases with the increase in powder feed rate up to a critical value. Beyond this critical value, the increase in deposition rate is more definite and the cross-section of track is similar to a circle. This is because at a higher powder feed rate, the powder flux shields the laser beam from coming in contact with the substrate and causes insufficient melting of the substrate. This results in a cylindrical track geometry due to improper surface tension, wetting and enhanced powder catchment. However, this kind of track geometry is undesirable because of its poor adhesion with substrate [39].

3. *Scan speed*: The scan speed affects the laser energy deposited and powder fed per unit length, as it is directly related to interaction time. Interaction time is the time duration for which the laser beam dwells at any point during the processing and is defined as the ratio of laser spot diameter to scan speed. The variation in the scan speed affects the amount of energy and material fed into the melt pool, which affects the geometry and properties of the components.

Laser energy per unit length and material fed per unit length affect the track width and height. Transverse traverse index influences the fabrication of overlapped or porous structures.

However, the effect of basic parameters can be accounted for with the following two parameters:

$$\text{Laser energy per unit length} = \frac{P}{V} \qquad (2.2)$$

$$\text{Powder fed per unit traverse length} = \frac{m_p}{V} \qquad (2.3)$$

where, m_p stands for powder feed rate. The parameters 'laser energy per unit traverse length' and 'powder fed per unit traverse length' govern the laser energy and the material available for the single track deposition respectively.

At extremely high laser energy per unit traverse length and lower powder fed per unit traverse length, vaporisation of the feed material may occur. This results in very thin or no track formation. On the contrary, at extremely low laser energies per unit traverse length and higher powder fed per unit traverse length, fusion of fed material may not occur and discontinuous tracks are formed. Hence, a processing window is required to balance both the parameters. The relationship between the applied heat source power and the heat source velocity is a key parameter of PBF and PFD processes. This relationship is important for eliminating process-induced porosity and determining grain morphology. The parameters will be obtained, resulting in uniformly fused continuous tracks.

4. *Laser spot size*: Laser beam spot size refers to the diameter of the laser beam on the target. Laser power and spot size affect the intensity of the laser beam fed into the material.

5. *Transverse traverse index/degree of overlap*: When a large surface is to be covered or a porous layer is to be fabricated, a number of tracks are laid adjacent to one another with an overlap for covering large surfaces and without/limited overlap for porous layers. Hence, another parameter that plays a vital role during the formation of multitrack structures is the transverse traverse index. This can also be referred to as the overlap rate. A higher overlap rate results in good surface

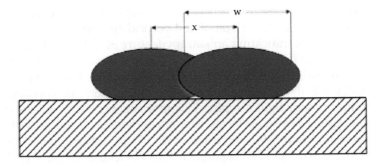

Figure 2.12 Schematic diagram of a transverse section of a transverse traverse index.

roughness, low porosity and well mechanical properties, while building time increases [40]. Figure 2.12 represents a schematic diagram of
a transverse section of a transverse traverse index.

$$\text{Transverse traverse index (i)} = \frac{x}{w} \qquad (2.4)$$

6. Scan strategy also has a significant effect on the properties of the components fabricated by LAM. The path that the heat source follows
during selective melting or deposition for lasers is classified as the
scan strategy. Scan strategies for PFD tend to be relatively simple as
a result of limitation imposed by movement of the powder or wire
feeding system. Unidirectional and bidirectional fills are both standard PBF processing techniques. These strategies use rectilinear infill
to melt a given part layer. Both unidirectional and bidirectional fills
are also used in LM systems. Island scan mode is a common strategy
in LM. The scan area is divided into many islands and within each
island a laser spot is scanned in one direction. The direction of scanning will be perpendicular for the island lying next to it. These islands
are melted selectively to evenly distribute the heating and reduce the
residual stresses in a random order. Following the selective melting of
the islands, the laser is scanned around the outer-contour of the slice
to refine the surface finish of the fabricated part [41]. The scan strategy has a direct impact on process parameters and thus, heat source
power and velocity must be optimized for a given scan strategy.
7. *Hatch angle*: The hatch angle θ refers to the angle between laser scanning directions on consecutive layers. For example, a hatch angle
of 90° means that direction of deposition of two consecutive layers
will be perpendicular to each other and the orientation of the fourth
melted row will be the same as the rows of the first layer [42]. The
studies on Inconel 625 by PFD revealed that the variation of tensile
properties of the material is within the limits considering overlap

error bar when the hatch pattern is changed [39]. However, tensile properties of both horizontal and vertical specimens vary with the build angles, indicating anisotropy feature of LM parts [40,43].

8. *Building direction*: Building direction refers to the acute angle between the longitudinal axis of sample and the vertical axis. The LM specimens built vertically have higher tensile strength and elongation as compared to horizontally built samples. This indicates significant anisotropy of LM parts. The average tensile strength for horizontal and vertical position is 624 and 669 Mpa, respectively. Thus, vertically built samples were 2.8% stronger as compared to that of other configurations [43].

9. *Layer thickness*: Layer thickness defines the thickness of each layer in the layer-by-layer fabrication of components by AM. It is found that layer thickness has minimal effect on the mechanical properties of the components.

2.7 Materials for LAM

The quality of the feed material determines the quality of the LAM built parts. The powder quality depends on the size and shape, composition, surface morphology and amount of internal porosity, and it determines physical variables, such as flowability and apparent density. Flowability is important for PBF and PFD, while apparent density is critical for PBF. Flowability refers to the ability of powder to flow while apparent density defines packing of a granular material. Spherical particles improve flowability and apparent density. Smooth particle surfaces are better than surfaces with satellites or other defects. Fine particles have better apparent density by filling the interstitial space between larger particles, but flowability may get reduced [30,44].

There are a variety of atomization techniques for generating metal powder, each delivering particular varieties in powder quality such as gas atomization (GA), rotary atomization (RA), plasma rotating electrode process (PREP), etc. Porosity in the powder feedstock is common for techniques such as gas-atomization (GA) that entrap inert gas during production. This entrapped gas is transferred to the part, during rapid solidification, and results in powder-induced porosity in the fabricated material. These pores are generally spherical and are formed as a result of the vapour pressure of the entrapped gas. PREP powder parts displayed insignificant porosity, while GA and RA powder parts both exhibited gas induced porosity [45].

PBF and PFD involve the complete melting of powders during the processing. The properties obtained by these processes are also comparable. PBF has been successfully deployed for processing pure metals such as Ti [46–48], Ta [49,50], Au [51], etc. Table 2.3 presents the comparison of the tensile mechanical properties of CP-Ti processed by different technologies [46].

PFD also involves the complete melting of feed material and is also deployed for processing pure metals. Research works have been done on the

Table 2.3 Comparison of the tensile mechanical properties of CP-Ti processed by different technologies

Processing method	Yield strength	Ultimate tensile strength	Fracture strain
PBF	555 ± 3	757 ± 12.5	19.5 ± 1.8
Sheet forming	280	345	20
Full annealed	432	561	14.7

Source: Attar, H. et al., *Mater. Sci. Eng. A*, 593, 170–177, 2014 [46].

development of porous structures for biomedical applications by Titanium [52], Tantalum [53]. PFD has significantly increased the flexibility in processing of complex-shaped three-dimensionally interconnected metallic implants with functionally graded porosities. Porosities and mechanical properties of laser-processed Titanium structures with interconnected porosity and novel design can be tailored by adjusting the process parameters of PFD. Young's modulus and 0.2% proof strength of porous Ti samples having porosity around 35–42 vol.% are close to those of human cortical bone.

LAM is also deployed for processing various metallic alloys. Ti-based alloys, typically Ti–6Al–4 V, are extensively used for aeronautical and medical applications [54–57] due to their exclusive chemical characteristics and mechanical features and biocompatibility. Inconel 625 and 718 are most commonly used nickel-based super alloys due to an enhanced balance of mechanical properties such as creep, damage tolerance, tensile properties and corrosion resistance for high-performance components of jet engines and gas turbines [39,58–60]. Aluminium alloys have been extensively applied in various sectors, such as automotive, aerospace, etc. This is because of the attractive combination of excellent weldability, high thermal conductivity and good corrosion resistance. But high reflectivity and high thermal conductivity of aluminium alloys limit the processing of aluminium alloys by a laser melting process compared with producing other metal powders such as titanium alloys, nickel-based alloys, stainless steel, etc. [61–65]. Tantalum-Titanium alloys are a promising choice for a high strength to density ratio. The choice of combining the materials holds good for biomedical applications mainly because of the high biocompatibility, corrosion resistance and good mechanical properties it possesses [66]. PBF and PFD technologies have been successfully deployed for fabricating stainless steel [67,68] and tool steel [69–71] components. Tool steel is known for its high-volume effectiveness in additive manufacturing of die-casting tooling [69], while stainless steel grades such as SS316, SS304, etc., have been widely used in aircraft, automotive and medical industries due to their high corrosion resistance, mechanical properties and high cost-effectiveness [40,67]. Table 2.4 presents various commercial systems with compatible metal alloys.

Table 2.4 Commercial systems with compatible metal alloys

Process	Commercial manufacturer		Materials
Powder fed deposition	OPTOMEC	Titanium	CPTi, Ti6-4, Ti6-2-4-2 Ti6-2-4-6, Ti48-2-2, Ti 22AI-23Nb
		Nickel	IN625, IN718, IN690, HastelloyX, Waspalloy, MarM247, Rene 142
		Stainless steel	13-8, 17-4, 304, 316, 410, 420, 15-5PH, AM355, 309, 416
		Tool steel	H13, S7, A-2
		Cobalt	Stellite 21
		Aluminum	4047
		Copper	Cu-Ni
	DMD	Fe based	4140 steel 4340 steel 300 Maraging H13 P20 P21 S 7 420SS 316L SS 304 SS 17-4 PH SS 15-5 PH SS CPM1V Invar
		Co based	Stellite 6 Stellite 21 Stellite 31 Stellite 706
		Nickel base	IN 718 IN 625 C-276 Nistelle C Wasp alloy
		Titanium base	CP Ti Ti-6Al-4V
		Aluminum base	4047 Al 6061 Al
		Cu base	Al-bronze Cu-Ni

(Continued)

Table 2.4 (Continued) Commercial systems with compatible metal alloys

Process	Commercial manufacturer		Materials
Powder bed fusion	EOS	Maraging steel	EOS Maraging Steel MS1
		Stainless steel	EOS Stainless Steel GP1
			EOS Stainless Steel PH1
			EOS Stainless Steel 316L
			EOS Stainless Steel CX
			EOS Stainless Steel 17-4PH
		Nickel alloy	EOS Nickel Alloy IN718
			EOS Nickel Alloy IN625
			EOS Nickel Alloy HX
		Cobalt chrome	EOS Cobalt Chrome MP1
			EOS Cobalt Chrome SP2
			EOS Cobalt Chrome RPD
		Titanium	EOS Titanium Ti64
			EOS Titanium Ti64ELI
		Aluminium	EOS Aluminium AlSi10Mg

Source: https://www.optomec.com; http://dm3dtech.com; https://www.eos.info [33–35].

2.8   Processing issues in LAM

LAM brought a revolution in the mainstream manufacturing sector mainly because of the freedom provided by the process in terms of design, material, energy control, etc. However, there are certain processing issues associated with LAM processes, such as delamination, cracking and residual stress, porosity, stair-stepping effect, balling and generation of fine features. Figure 2.13 presents the processing issues in LAM along with remedial measures.

2.8.1   Delamination, cracking and residual stress

Residual stress developed during processing leads to delamination and cracks, and it prohibits the usage of many metals in LAM. High thermal stress involved in the LM process leads to this residual stress and results in cracking and delamination of the parts. These stresses should be reduced to produce dense and quality parts. There are mainly two mechanisms that induce residual stress.

1. Stress induced in the solid substrate under the layer undergoing melting
2. Stress due to cooling of melted layers on top

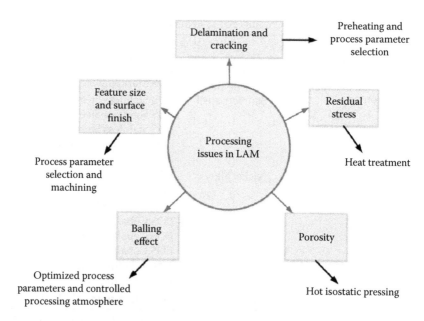

Figure 2.13 Processing issues and remedial measures in LAM.

In the first mechanism, if the temperature of the upper layer is larger than the solidified layers it will try to expand and will be restricted by the lower solidified layers. This results in compressive stress, which may exceed the yield strength of the upper layers. This leads to plastic deformation of the upper layers, and this stress state is converted to tensile stress when layers cool down. These tensile stresses act as residual stress and induce cracking on the parts when they exceed the ultimate tensile strength locally. In the second mechanism, shrinkage of the melted layer on top due to thermal contraction is inhibited by the lower layers, which results in tensile stress and compressive stress in the upper and lower layers, respectively. Lowering the thermal gradient reduces the thermal stresses and hereby the amount of cracking. Preheating the base plate reduces the temperature gradient and lowers the rate of cooling, which converges to reduced distortion in the crystal structure. Remelting of the layers was not identified as an ideal option for controlling cracking and delamination since it favours the distortion in the crystal structure [72].

2.8.2 Porosity

Porosity is common in LAM systems and can be of mainly two types: Lack of fusion porosity and porosity due to gases. Lack of fusion porosity is mainly due to insufficient melting of the powder material and is usually irregular in shape with elongation along the plane of the layer. Porosity due to gases can be due to entrapment of gases from a powder system or releases of gases inside the powder particles. These are usually spherical in nature and are not location specific like the other one. Gas porosity due to gas entrapment in powder was discussed before in this chapter. It is observed that the amount of gas porosity is low as compared to lack of fusion porosity. Decreasing the laser power and increasing the scan speed both results in an increased porosity [73].

Porosity due to lack of proper fusion is influenced by the processing parameters. Thus, a proper processing window is required to process materials with minimal porosities. Laser power, scan speed and hatch spacing influence the fusion of materials and thereby affect the porosity of the components. It is observed by researchers that the influence of laser power on the porosity is more than scan speed. Reduction of laser power and increase in scan speed have a similar effect on the porosity as both result in the reduction of laser energy and melt pool, which leads to the formation of pores. In addition to this, an increase in hatch spacing increases the porosity as it leads to insufficient overlap between scan tracks and results in incomplete consolidation. Thus, a processing window should be developed for proper control of fusion and porosity [74]. Scientists at Lawrence Livermore National Laboratory, US, discovered the reason for porosity in PBF processes. According to them, the gas flow due to evaporation when the laser irradiates the metal powders clears away powder near the laser path, and this phenomenon reduces the powder available for the next pass of laser. This results in tiny gaps and defects in the fabricated component [75].

2.8.3 Balling effect

Balling effect is associated with LAM and refers to a complex physical metallurgical process [55] influenced by material characteristics and processing conditions. During the process, the molten track has a tendency for shrinkage to reduce the surface energy due to surface tension. This leads to the formation of balling effect and hinders the quality of the part produced. Balling effect increases the surface roughness and porosity of the part fabricated. At the same time, it hinders the movement of the blade in PBF processes.

The influencing factors of the balling phenomenon are oxygen content, laser power, scan speed, scan interval, etc. [77]. The increase of oxygen in the processing chamber enhances oxidation of the molten pool and reduced wetting with oxide at the wetting interface. Processing parameters such as laser power and scanning speed should be properly selected for ensuring proper wettability. High scan speed and low laser energy reduce the molten pool dimension and the substrate melt pool contact area. They eventually lead to unfavourable wetting of the substrate, spreading and flowing. This results in the balling phenomenon. Thus, the phenomenon can be controlled by adjusting the process parameters. The balling phenomenon can also be lightened by remelting of tracks, through which the metal balls could be melted and then wet the surface favourably [76,77].

2.8.4   Stair-stepping effect

The stair-stepping effect is another common limitation in the AM process due to layer-by-layer manufacturing methodology. This hampers the surface quality of the parts along the built direction especially for inclined and curved surfaces. Increasing layer thickness has a significant effect on poor surface finish of the parts. The stair-stepping effect is quantified using *Total Waviness* of the parts. The selection of proper build direction and process parameters is important for reducing the stair-stepping effect [78]. Figure 2.14 presents the stair-steeping effect in LAM.

2.8.5   Feature size and surface finish

Feature size and surface finish are concerns for AM users although the required surface finish can be achieved by post-processing, which will be explained later on. Feature size depends on the size of the energy source. Laser being the energy source in LAM, the feature size depends on the diameter of the laser. It can be considered that PBF systems can fabricate parts with reduced feature size as compared to PFD systems due to the inertia involved. The smaller feature size can be obtained at the expense of larger built time.

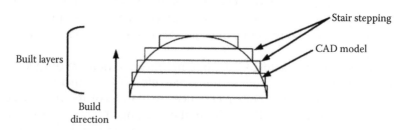

Figure 2.14 Stair-stepping effect.

2.9 Post-processing of laser additive manufacturing parts

The bulk and surface properties as well as geometry achieved in LAM-built material are influenced by distortions due to heating, partially melted powders, solidified melt droplets and surface variations brought by the laser movement and processing strategy. As-deposited surface roughness, dimensional accuracy and properties are inadequate for many industrial applications. Therefore, a certain amount of post-processing is required for LAM components [79].

The first stage of post-processing is soon after the part is built. In the PBF process, the fabricated component will be enclosed by powder material and this loose or unfused powder is removed during post-processing. In PBF, support structures may be generated based on the geometry and are to be removed by cutting. In addition to this, PBF and PFD processes require removal of part from the substrate. This is usually done with wire electrical discharge machining or a saw.

The next stage of post-processing depends on the application. This can be surface finish enhancement, aesthetic enhancements, property enhancements and so on. Surface finish enhancements have been done using conventional subtractive manufacturing techniques such as computer numerical control (CNC) milling and polishing, glass blasting, or ultrasonic machining [80]. Recently, post-processing of LAM components were performed using laser-based techniques such as laser shock peening (LSP), laser finishing and laser annealing (LA). The components built using LAM are inherited with tensile residual stress on the surface due to rapid heating and cooling during material processing. Thus, LSP, an advanced surface engineering process, introduces compressive residual stresses into materials, thereby improving product life through increased resistance to many surface-related failures, such as wear and corrosion. LSP on LAM fabricated IN718 showed significant improvement in mechanical properties such as hardness, corrosion and wear resistance by 27%, 70% and 77%, respectively [81]. LSP studies on LAM fabricated NiTi and TiNiCu-based shape memory alloys showed changes in the surface morphology such as increment in surface roughness and decrement in lots of peak structures. The microstructures were closely bonded, and there were no cracks formed on the surface due to LSP. The X-ray diffraction (XRD) graphs showed amorphization of the samples and broader peak width is observed after LSP. Differential Scanning Calorimetry (DSC) graphs also show that compositions were capable of producing peaks, in both a heating and cooling curve. Further, LA was performed to overcome the problem of amorphization and recrystallize the formed samples. The increase in grain size of all samples was visible with Atomic Force Microscopy (AFM) and XRD results. The DSC graphs showed peaks formed in them after LA. Thus, combination of LSP and LA is an exceptional combination of two advanced techniques for post-processing of LAM components [82]. Femto-second laser was employed for removing extra material, reducing surface roughness and improving the geometrical quality of complex micro-scale topology features, such as holes with a higher aspect ratio. A significant reduction in the surface roughness was observed. It can be concluded that femtosecond laser radiation is suitable for post-processing of thermal sensitive parts with complex features due to high spatial sensitivity and controlled energy input [79].

Post-processing heat treatment is also applied for LAM components to adopt the properties of the components to the working conditions or to reduce the thermal stress induced [83]. Thus, the desired microstructure and mechanical properties for service conditions can be achieved by various heat treatment procedures. These treatments alter the grain size, grain orientation, porosity and mechanical properties. Relieving of internal stress is another aspect associated with heat treatment. As discussed earlier, LAM components have residual internal stress due to a high thermal gradient. Thus, annealing is performed on LAM components for reducing the internal residual stresses [84]. Solution treatment and ageing procedures are common for precipitation-hardened materials, such as nickel-based super alloys. Solution treatment helps in dissolving the undesirable phases, while aging enables the formation and growth of precipitation phases. These processes are usually done sequentially. The processing conditions and time for solution treatment should be properly selected for dissolving the precipitates. Once the solution treatment is done, aging is carried out to increase the hardness of the material [30]. The standard heat treatment procedures for Inconel 718 are as follows:

1. Solution treatment (980°C, 1 h/air cooling) and double aging (720°C, 8 h/furnace cooling at 55°C/h to 620°C, 8 h/air cooling)
2. Homogenisation treatment (1080°C, 1.5 h/air cooling) and solution treatment (980°C, 1 h/air cooling) and double aging (720°C, 8 h/furnace cooling at 55°C/h to 620°C, 8 h/air cooling) [85]

The hot isostatic pressing (HIP) process has been extensively used in the healing defects such as cavities, voids and hot cracking. It was reported that HIP had a significant effect on the mechanical properties and microstructure in several research works [86]. Enhancement and reduction of the scatter in the mechanical properties was observed as compared to an untreated sample. This can be attributed to the ability of HIP to heal the bulk and surface defects. Since HIP involves high pressure and temperature, fracture surfaces of the cracks were closed mechanically by the high temperature creep, then bonded together, and finally diffusion homogenised.

Table 2.5 Mechanical properties of sintered and infiltrated material

Material	Tensile strength (MPa)	Rockwell hardness
Sintered	96–105	25.2
45 min infiltrated material	92–102	89.6
90 min infiltrated material	132	69.5

Source: Sindel, M. et al., Direct selective laser sintering of metals and metal melt infiltration for near net shape fabrication of components, *Proceedings of Solid Freeform Fabrication Symposium*, 94–101, 1994 [89].

Property enhancements can be done with non-thermal techniques, such as shot peening. Shot peening is a mechanical surface treatment technique in which small balls are impacted on the surface of a component. The repeated impacts of the balls induce compressive residual stress and refine the microstructure. This helps in delaying the crack initiation and hinders the crack propagation [87]. Thus, the mechanical properties and microstructure can be tailored as per the requirement by shot peening. Infiltration is another post-processing technique used in laser-sintered components. Porous LS part is heated in contact with the infiltrant to a temperature at which the infiltrant is molten and will soak into the part through capillary action. The infiltrant solidifies on cooling and produces the final part. The significant attention is on the ability of the infiltrant to wet the solid preform and form the dense solid [88]. The strength of the structure after infiltration is a function of the time period of infiltration as shown in Table 2.5.

It was observed that the tensile strength after 45 min infiltration did not improve much as compared to sintered structure. This is due to the presence of singular large pores in the infiltrated material. These pores are filled slowly since capillary pressure is inversely related to the radius of the capillary. Thus, prolonged infiltration is required to enhance the strength of the components [89].

2.10 Benchmarking of laser additive manufacturing process

LAM process is growing, and the biggest challenge for a wider industrial acceptance still stands as the need for more reliable, repeatable and precise machines for additive manufacturing. Thus, a comprehensive benchmarking study for the selection of an additive manufacturing machine for LAM is necessary. Benchmarking examines different LAM systems with regard to conditions that become necessary for manufacturing, such as accuracy, material, mechanical properties, speed and reliability. A benchmark model is used to test these conditions and to check the process limitations. In addition to this, it can be used to optimise each process iteratively. Dimensional analysis, surface roughness analysis and mechanical properties such as density, hardness, strength and stiffness were analysed. Dimensions of the benchmark model are 50 × 50 × 9 mm. The benchmark model consists of a sloping plane and rounded corner, which can verify the stair-step. It also has a 2 mm thin plane that allows the study of thermal distortions and warpage. Small holes and cylinders (0.5–5 mm diameters) and with thin walls (0.25–1 mm thickness) can be used for evaluating precision and resolution. The sharp edges of the model help to study the influence of heat accumulation and possible scanning errors. The overhanging surfaces can demonstrate the capability to manufacture overhanging structures without support structures. All of the geometrical features help to analyse the accuracy of the process in x, y and z-direction. Figure of the model used by Kruth et al. is presented in the published literature [90]. In the model by Yasa et al., different equipment providers for LAM for aero engine part manufacturing using Inconel 625 powder have been used for comparison of machines. The factors such as dimensional accuracy, surface quality, geometrical resolution, necessity of support structures, density, hardness and process limits were addressed. Figure of the model used by Yasa et al. is presented in the published literature [91].

Geometrical accuracy is studied in an article by A.L. Cooke and J.A. Soons [92]. In this study, the benchmark pieces are fabricated using electron and laser-based additive manufacturing. The test part is the Aerospace Industries Association (AIA), National Aerospace Standard, NAS 979 standardised test piece. The test piece consists of circle-diamond-square features, with an inverted cone. The figure is shown in published literature [92]. Vandenbroucke and Kruth [93] manufactured the benchmark pieces to estimate the aptness of LAM components in use in the medical field. The benchmark pieces shown in the figure published in published literature [93] are designed for evaluating the accuracy of the LAM process and the suitability to manufacture small details. Surface roughness of manufactured pieces were also studied. In this study, mechanical properties, such as density, hardness and tensile strength of LAM components were compared with bulk materials mechanical properties.

2.11   Laser additive manufacturing at RRCAT

Realizing the importance of LAM and allied technologies, a comprehensive research and development programme was initiated in the year 2003 at Raja Ramanna Centre for Advanced Technology (RRCAT). A number of components for engineering and prosthetic applications have been manufactured using the LAM system integrated at the RRCAT's LAM Laboratory. A detailed description of some of the parts built at LAML is presented in our earlier published work [27].

2.11.1 Laser additive manufacturing system at RRCAT

The schematic arrangement of the system is shown in Figure 2.15. It essentially consists of a 2 kW fibre laser system, a 5-axis workstation in a glove box, a computerised numerical controller, a coaxial nozzle, a twin powder feeder, gas analysers and thermal imaging camera. In the system, a commercially available 2 kW fibre with an output power range of 50–2050 W and emission wavelength of 1080 nm is used. The system possesses a switching on/off time of 80 μs and an output modulation rate of 5 kHz. The laser beam is randomly polarized and has beam product parameter less than three. The system has a delivery fibre with a core diameter of 50 μm, and the output laser beam is passed through refractive focusing optics and an indigenously developed compact coaxial nozzle is employed for laser additive manufacturing.

The system is integrated with a 5-axis workstation for job manipulation. The five axes are X, Y, Z, V and W. The X, Y and Z axes are linear traverse axes mutually perpendicular to each other, whereas V is a tilt axis about the Y-axis, for tilting the laser head, and W is a continuous rotational axis about the X-axis. The effective stroke length of the linear axis is 250 mm. The angular tilt of the V axis is ±110°, whereas the W axis is capable of a 360° continuous rotation. The manipulator is interfaced with a standard computer numerical controller for manipulating the workstation. The laser head coupled with a manipulator is mounted in a glove box. The glove box is essentially required for maintaining controlled atmospheric conditions during processing. Oxygen and moisture are the main impurities in the atmosphere, which affects the properties

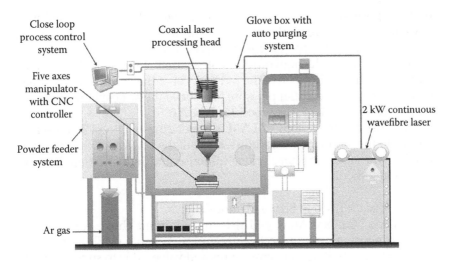

Figure 2.15 LAM system at RRCAT.

Figure 2.16 In process LAM at RRCAT.

of the deposited bulk materials. Therefore, the system is integrated with oxygen and moisture analysers. The desired purity levels are achieved by purging high purity grade Argon gas. In case there is an increase in the impurity level, the high purging rate is used to reinstate the indented purity level in the glove box. The purity level of the glove box is retained by keeping the differential pressure just above the atmospheric pressure. Figure 2.16 shows the LAM process at RRCAT.

The list of applications of LAM is appending due to global research efforts. In the following case studies, some of the LAM applications developed at our laboratory for various in-house and industrial applications are briefly described.

Case Study 1: LAM of Shape Memory Alloys
Shape Memory Alloy (SMA) is an important choice for manufacturing many micro-electro-mechanical systems (MEMS) products, such as actuators in micro-pumps and valves, due to super elasticity and thermal shape memory effects. SMA can deform its shape in a

controlled way, and this is possible because of the presence of two phases it possesses. Phase transformations between a lower temperature martensite phase and higher temperature austenite phase result in the properties of these materials. Although different materials have been tried for shape memory effect, Ni-Ti is preferred due to its ability to retain the properties for a longer period of time. Although several routes have been employed for the fabrication of SMA, LAM has the unique advantage of fabricating complex structures and inherent flexibility in process control. In the current study, a 2 kW fibre laser-based additive manufacturing (LAM) system at RRCAT is deployed for laser additive manufacturing of various deposits involving Ni and Ti. Ni and Ti compositions with powders in three different weight percentages (Ni-45% Ti-55%, Ni-50% Ti-50% and Ni-55% Ti-45%). SEM, XRD, DSC and mechanical properties are studied to understand the variation in characteristics for the three conditions. Figure 2.17 shows the variation of hardness among the different compositions. The average microhardness is more for Ti-rich samples due to the formation of hard phases caused by the increased Ti content. Since the microhardness is more for NiTi55 and NiTi45, it may be concluded that the ductility is lower for these compositions [94].

A scanning electron microscopy study of the three compositions reveals sporosity in NiTi55 and NiTi45, while nonporous structures are developed using NiTi50. Martensite peaks are not seen in NiTi55 and NiTi45. Thus, it is concluded from all the results that NiTi50 is the best combination for shape memory characteristics. To improve

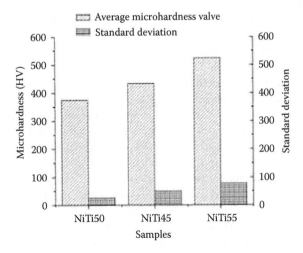

Figure 2.17 Microhardness for different compositions. (From Shiva, S. et al., *Opt. Laser Technol.*, 69, 44–51, 2015 [94].)

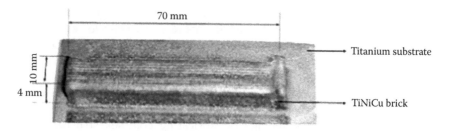

Figure 2.18 TiNiCu brick developed by LAM. (From Shiva, S. et al., *J. Mater. Process. Technol.*, 238, 142–151, 2016 [95].)

the ductility and reduce the porosity of the product to be formed, copper is introduced to Ni–Ti alloys. Three compositions of TiNiCu (Ti-50%Ni-45%Cu-5%, Ti-50%Ni-35%Cu-15%, Ti-50%Ni-25%Cu-25%) are selected for the investigation [95]. Figure 2.18 presents the TiNiCu brick developed by LAM.

Addition of copper along with Ni-Ti enhances the phase transformation property of the SMA. SEM studies reveal that TiNiCu5 does not carry pores, cracks and other defects. Microstructures observed are fine in nature. The XRD results reveal both austenite and martensite phases in all three compositions. The heating and cooling curves in DSC graphs establish the enhancement in phase transformation with Cu. TiNiCu15 and TiNiCu25 are brittle while comparing to TiNiCu5 as the transformation curves are not as deep as TiNiCu5. This corelates with the results of microhardness tests presented in Figure 2.19. It is concluded from the testing that, TiNiCu5 is the best among the samples under investigations [82].

Post-processing of bulk SMA fabricated by LAM was subjected to laser annealing [96] and laser shock peening to reduce the brittleness and strength of the material [82].

Case Study 2: LAM of SS-Ti Transition Joint

As the diverse and extreme operating conditions of today's industrial arena cannot be satisfied by single-material components, the manufacturing community is challenged to search for better techno-economical solutions for fabricating advanced components made of dissimilar materials, such as Stainless Steel-Titanium (SS-Ti). Joining of SS-Ti using conventional welding processes does not yield sound and long-lasting joints due to metallurgical incompatibility. Therefore, the traditional way of joining these materials is roll bonding, pressure welding, explosive welding, vacuum brazing and diffusion bonding. When these joints are exposed to cyclic loading conditions, they are more prone to premature interfacial failure due to a mismatch in thermal physical properties. Figure 2.20 presents the SS-Ti transition joint built at RRCAT.

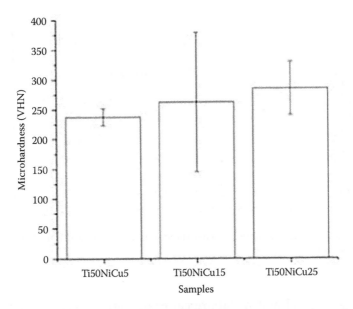

Figure 2.19 Microhardness results of TiNiCu samples. (From Shiva, S. et al., *J. Mater. Process. Technol.*, 238, 142–151, 2016 [95].)

Figure 2.20 LAM SS-Ti transition joint.

An in-house integrated 2 kW fibre laser-based LAM system with processing window is identified as laser power in the range of 600–1200 W, powder feed rate in the range of 3–6 g/min, and scan speed in the range of 300–500 mm/min. Laser additive manufactured SS-Ti transition joints are tested for various

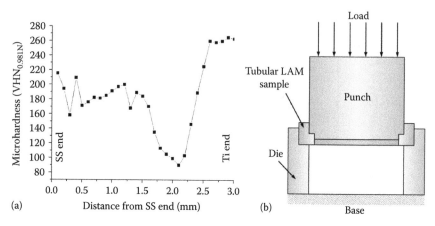

Figure 2.21 (a) Microhardness measurement of SS-Ti joint and (b) shear strength testing setup.

mechanical and metallurgical examinations. Figure 2.21a presents the results of microhardness measurement across the cross section of the SS-Ti transition joint. The transition joints are also subjected to shear strength testing using the setup shown in Figure 2.21b. The shear strength of these joints is found in the range of 335–400 MPa.

Case Study 3: Laser Additive Manufacturing of Honeycomb Geometry Orifice Plates

Liquid sodium, which is a coolant, enters the hot pool of the fast breeder reactor from the subassemblies at a temperature of 820 K. The temperature of sodium is maintained by judicious allocation of flow through the various zones of the core. This stringent requirement of flow allocation is achieved by permanent installation of a stack of honeycomb geometry orifice plates. This honeycomb geometry orifice plate has a complex structure with several hubs and ribs joining each other. It is 120° symmetric, and 60° rotation gives a full offset with respect to the preceding plates stacked together. The desired flow allocation using these orifice plates is achieved by varying the number and orientation of plates in the stack. Several techniques (such as electric discharge machining, conventional casting, investment casting, etc) are tried to manufacture these orifice plates. A limited success is achieved and further improvements are required. The development of these honeycomb geometry orifice plates is undertaken by laser additive manufacturing (LAM) with the input received from our downstream partner at IGCAR, Kalpakkam.

Figure 2.22 LAM of honeycomb geometry orifice plates.

The major challenge in fabrication of this orifice plate is its complex shape and hanging structure. Special methodology to tame these issues for the LAM of these orifice plates is developed. LAM of these orifice plates were carried out in two steps. In the first step, the first half is manufactured with a suitable cutting allowance. The deposited structure is cut along the diameter using an electric discharge wire-cut machine and subsequently, the other half is fabricated by keeping the deposited structure in an inverted position on the specially designed fixture. The fabricated orifice plates are evaluated for their shape, size and surface finish. It was found that the shape, size and surface finish of the LAM orifice plate is within requisite tolerances. Figure 2.22 gives a typical picture of honeycomb geometry orifice plates made by LAM at RRCAT.

Case Study 4: Hardfacing of Ni-based Tribaloy-T700 alloy on austenitic stainless steel 316L

LAM is employed for hardfacing to produce coating with one or more layers that protect the underlying substrate against wear, abrasion and corrosion and to improve the service life and economy of the components. Hardfacing of austenitic stainless steels is generally carried out with cobalt and nickel-based alloys, having resistance to both wear and corrosion over a wide range of temperatures and environments. Hardfacing alloys containing cobalt and boron are not suitable for nuclear applications as they transmute to undesirable isotopes or capture excessive quantities of neutrons in neutron flux environments. In addition, the increasing price of cobalt is attractive as per economics for use of cobalt-free hardfacing materials. Thus, Tribaloy 700 (T-700), a cobalt-free Ni-based hardfacing alloy, is studied. Three compositions as 100% T-700 as (1:0), 75%T-700-25% SS316L as (3:1) and 50% T-700-50% SS316L as (1:1) are selected to assure the deposited layer should be uniform and defects free, such as cracks, void, porosity, etc., with an excellent metallurgical bonding with the substrate of austenitic stainless steels 316L.

T-700 is deposited on the substrate of SS316L to increase the hardness of the stainless steel without affecting the bulk properties. Initially, a number of single tracks are deposited at various combinations of laser power (800–1800 W), scan speed (300–900 mm/min) and powder feed rate (3–12 g/min) for optimization of process parameters. The optimized processed parameters are Laser power – 1500 W, Powder feed rate – 6 g/min and Scan speed – 0.5 m/min with 2.5 mm laser spot size and maintaining 10 mm standoff distance with substrate. The single and multilayer overlap tracks are deposited with three combinations T700, T700-25SS and T700-50SS with 50% overlap of track.

It was observed that 100% T-700 deposition on SS316L revealed cracks along the longitudinal as well as transverse direction of deposits. These cracks are disappearing when the composition of T-700 is reduced from the deposited material. This crack formation can be attributed to the difference in thermomechanical properties of the materials. The clad layer for the 100% T-700 has a higher hardness value but cracks are observed in the deposition, which makes it not suitable for mechanical applications. The clad layer for T700-25SS as 3:1 mixing ratio has a hardness value around 2.5 times the substrate material SS316L and is considered as the best option of coating for industrial applications. Figure 2.23 presents the LAM deposited T-700 on SS316L and microhardness profile of different mixing ratios of T-700 and SS316L.

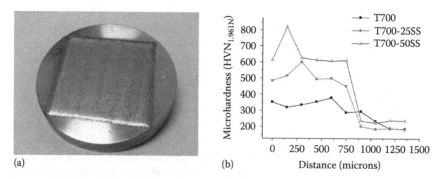

(a)   (b)

Figure 2.23 (a) LAM deposited T-700 on SS316L and (b) Microhardness profile of different mixing ratio of T-700 and SS316L.

Case Study 5: LAM of Hip Implant for Biomedical Applications

There is variation in shape and size of the human body parts. Moreover, the mechanical properties of natural hard tissues of the human body are different and depend on human age, origin, sex, food habits and lifestyle. Hence, prosthetics need customisation to meet the vivid requirements. LAM, being a technology for mass customisation, can be deployed to address the issue. Keeping this in mind, LAM laboratory at RRCAT starting working in this area and a prototype hip-implant is fabricated using LAM. It started with making of a solid model. Solid model of Ti-femoral stem with stem length 125 mm and proximal widths at mediolateral and anterior-posterior planes of 19 and 29 mm is drawn. The tool paths are generated using CAD/CAM software UG NX 6.0 and they are modified into customized G&M code. Subsequently, customized codes are transferred to the CNC controller of a 2 kW fibre laser-based LAM system. The paths are simulated and checked for geometry profile on a Siemens 810D controller. At the laser processing end, the desired controlled argon atmosphere ($O_2 < 25$ ppm and $H_2O < 35$ ppm) is achieved and test trials for the optimization of processing parameters are carried out. The optimum process parameters are obtained and defect-free tracks are deposited at laser power: 180–200 W; scan speed: 0.2–0.3 m/min; and powder feed rate 1.75–2.0 g/min. Subsequently, the femoral stem of the hip implant is laser additive manufactured at these optimum parameters. The fabrication of femoral stem is carried out in two steps. In the first step, the first half is manufactured with a cutting allowance of 3 mm and supporting structures (Figure 2.24a). The deposited structure is cut longitudinally using a wire-cut electric discharge machine and subsequently, the other half is laser additive manufactured by keeping the deposited structure in an inverted position on the specially designed fixture. Figure 2.24b presents the laser additive manufactured hip implant.

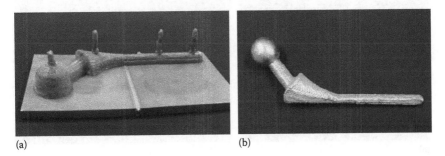

(a)   (b)

Figure 2.24 (a) Implant with supporting structure after LAM (b) Laser additive manufactured hip-implant.

Case Study 6: LAM of Dolero-50 Bushes

The bushes of a transfer arm gripper subassembly of a fast breeder reactor require high temperature wear and corrosion resistance properties. These can be met using nickel-based alloys, such as Colmonoy-6, Deloro-50, etc. The bushes have an 18 mm outside diameter, 15 mm inside diameter, and 20.2 mm height with a surface finish of 0.8 microns Ra value. In the past, these bushes have been made using conventional powder metallurgy and tungsten inert gas welding-based material deposition and subsequent machining route. This process has two major disadvantages: (1) it has a very narrow window for process parameters to get crack-free material deposition; (2) it involves multiple subtractive steps to get an actual component. It motivated us to investigate the fabrication of these bushes using Laser Additive Manufacturing as an alternative route. A 2 kW fibre laser-based Additive Manufacturing (LAM) has been deployed for the fabrication of these bushes. A sand bath at 400°C is used for preheating of the substrate. The process parameters used for the fabrication of these bushes are Laser Power – 1 kW, Powder feed rate – 4 g/min and Scan speed – 0.5 m/min. The fabricated bushes are buried in sand of a sand bath and slow natural cooling is allowed to avoid thermal cracking. The microhardness measurements indicate the hardness of 715–842 VHN in LAM deposited Deloro-50. Subsequent to LAM of Deloro-50 bushes, these are machined using a conventional machining process to meet the tolerance and surface finish requirements. A HSS tool with 10% Co, manufactured by powder metallurgy, is used for rough machining work and to get regular surface. Pre-final cutting is carried out using CERMET (Ni- bonded tungsten carbide) inserts for faster material removal rates, while final finishing is carried out using CBN inserts. This has resulted in reduction in material wastage and machining of hard materials. Figure 2.25 presents LAM built bushes in as built and post-machining condition.

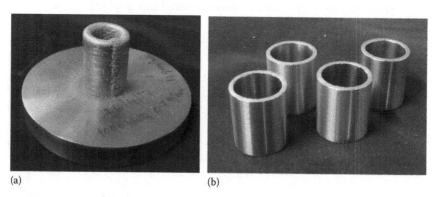

(a)   (b)

Figure 2.25 LAM built Dolero-50 bushes (a) as-built (b) after machining.

2.12   Conclusion

Additive manufacturing, often known as 3D printing, is changing the way in which things are being made and is revolutionizing future industrial production. Additive Manufacturing is one of the thrusts for the beginning of fourth industrial revolution with breakthroughs in emerging technologies. Manufacturers are eyeing supply chain reduction, reducing shipping costs and minimizing lead times through additive technologies. Among the segment of new generation AM processes that deliver metallic components (also called Metal Additive Manufacturing [MAM] process) using lasers, electron and plasma, lasers dominate MAM techniques with their unique characteristics and have been deployed commercially for fabricating complex and customized metallic components through Laser Additive Manufacturing (LAM). The process parameters play a key role in both the processes for fabricating optimized components with quality, accuracy and required mechanical properties. Various materials can be processed using LAM, which includes pure metals, alloys and ceramics. Post-processing steps such as laser annealing, laser finishing and laser peening are necessary for the final end usage of LAM components.

In engineering, the key economic benefits of AM originate from the redesigning of components based on the technology. Traditionally, freedom of designers is restricted by the manufacturability of a component from the direction of equivalent manufacturing and subtractive manufacturing. However, with the development of additive manufacturing technologies, many product and equipment structures can be reassessed from its viewpoint. Product and equipment design needs to focus more on functional optimization and less on manufacturability limitations, and 3D printing will make this possible. Topologically optimized components can be fabricated, which can hardly or not be realized with conventional methods. The same is true, for example, for functional parts such as heat exchangers or medical implants. AM can use almost *any* material to fabricate *any* complex part, in *any* quantity and in *any* location, for *any* industrial field. These five 'any's are the main advantages of 3D printing. Interdisciplinary research will tremendously accelerate the development of this technology. Two out of the seven additive manufacturing processes use laser for fabricating metallic components. Lasers with power in the range of 1–6 kW are currently deployed for LAM. With the advent of technology, a more powerful laser with better efficiency and process control can be used to solve some of the existing weaknesses of LAM technologies such as repeatability, reproducibility, layer thickness, deposition rate, etc. Looking ahead, it is essential to combine the additive technologies with other manufacturing techniques such as subtractive or formative manufacturing to improve the surface quality of the components. LAM combined with other manufacturing techniques forms hybrid laser additive manufacturing, which can resolve issues such as geometrical accuracy, cycle time and tailored properties. It is just the beginning of innovations in manufacturing through the deployment of LAM, and much more is hiding in the womb of the future to come.

Acknowledgements

The authors express their sincere gratitude to Dr. P. A. Naik, Director of the Raja Ramanna Centre for Advanced Technology (RRCAT), for his constant support and encouragement. Thanks are due to our collaborators Dr. I. A. Palani and Dr. S. Shiva of the Indian Institute of Technology, Indore, India; Dr. A. Sabreeswaran of Sree Chitra Tirunal Institute of Medical Science and Technology, Thiruvananthapuram, India; and Dr. C. Sudha of Indira Gandhi Centre for Atomic Research, Kalpakkam, India. The technical support of Mr. P. Bhargava, Dr. Atul Kumar, Mr. Harish Kumar, Mr. M. O. Ittoop, Mr. S. K. Mishra, Mr. C. H. Prem Singh, Mr. Dinesh Nagpure, Mr. C. S. Mandaloi, Mr. Upendra Kumar, Mr. Bhupendra Singh, Mr. Anil Adbol, Mr. Deepak Tampboli, Mr. Sokhen Tudu, Mr. S. D. Sharma, Mr. R. K. Gupta, Mr. P. Chinna Rao, Mr. D. K. Jain, Mr. G. Mundra and Mr. Saurav Kumar Nayak of RRCAT is thankfully acknowledged.

Theoretical questions

1. Define additive manufacturing and explain the various steps involved in additive manufacturing.
2. Why is the laser considered an effective energy source for material processing?
3. What are the different freedoms offered by LAM?
4. Define LAM and explain various LAM processes, stating the similarities and differences.
5. What are the essential components required for a LAM system?
6. How can coaxial material feeding be superior over lateral feeding?
7. Explain the powder feeding mechanisms in different LAM systems.
8. Explain the powder levelling mechanism in PBF.
9. Why is the F-θ lens made with built-in distortion?
10. Explain the function of VarioSCAN and galvanoscanner.
11. How does the hatch angle and scanning strategy affect properties of LAM built components?
12. What are the issues in LAM of aluminium and its alloys?
13. What are the major applications and properties of the following common LAM alloys?
    a. Inconel
    b. Titanium and its alloys

    c. Stainless steel

    d. Tool steel

14. What are the primary reasons for delamination and porosity in LAM components? How can it be rectified?

15. What is the balling effect in LAM and how can it be prevented?

16. Explain any three advanced post-processing techniques for LAM components.

17. Why is benchmarking required for LAM? How can it be performed?

18. Why is LAM of a stainless steel-titanium transition joint significant?

19. Explain hardfacing with the help of an example.

20. How can LAM contribute to biomedical engineering?

Numerical questions

1. LAM of a material 'X' was performed at a laser power of 1 kW, scan speed of 0.5 m/min and powder feed rate of 5 g/min. Evaluate the laser energy per unit length and powder fed per unit length.

   Solution:

   $$\text{Laser power } (p) = 1 \text{ kW} = 1000 \text{ W}$$

   $$\text{Scanning speed } (v) \ 0.5 \ \text{m/min} = 0.008 \ \text{m/sec}$$

   $$\text{Powder feed rate } (f) = 5 \ \text{g/min} = 0.083 \ \text{g/sec}$$

   $$\text{Laser energy per unit length} = \frac{P}{V} = \frac{1000}{0.083} = 1.25 \times 10^5 \ \text{J/m}$$

   $$\text{Powder Fed per unit length} = \frac{f}{V} = \frac{5}{0.5} = 10 \ \text{g/m}$$

2. LAM of a material 'A' requires minimum laser energy per unit length of 1.5×10^5 J/m. Find the minimum power required if the minimum scan speed is limited to 0.2 m/min.

   Solution:

   $$\text{Laser energy per unit length} = 150{,}000 \ \text{J/m}$$

Scanning speed (v) 0.2 m/min = 0.0033 m/sec

$$\text{Laser energy per unit length} = \frac{P}{V} \Rightarrow P = 500\,W$$

Unsolved numerical questions

1. LAM of a material Y was performed at a laser power of 2 kW, scan speed of 0.2 m/min and powder feed rate of 10 g/min. Evaluate the laser energy per unit length (J/m) and powder fed per unit length (g/m).
2. LAM of a material Z requires minimum laser energy per unit length of 150,000 J/m and powder fed per unit length of 20 g/min. Find the minimum laser power (W) and powder feed rate (g/min) required if the minimum scan speed is limited to 0.3 m/min.

References

1. Standard terminology for additive manufacturing technologies. ASTM International ASTM F2792-10e1, West Conshohocken, PA.
2. Steen, W., Mazumder, J. 2010. *Laser Material Processing* (4th ed.). London, UK: Springer-Verlag.
3. Paul, C.P., Kumar, A., Bhargava, P., Kukreja, L.M. 2013. Laser assisted manufacturing: Fundamentals, current scenario and future applications. In: J. P. Davim (editor-in-chief), *Non Traditional Machining Processes*, London, UK: Springer-Verlag, pp. 1–34.
4. 3D printing market to grow by 23%. 2016. *Metal Powder Report*, 71:470–471.
5. EOS completes pilot phase for 3D printing process monitoring system. 2016. *Metal Powder Report*, 71:471.
6. Siemens acquires AM specialist. 2016. *Metal Powder Report*, 71:469.
7. 3D Printer and 3D Printing News. http://www.3ders.org (Accessed on October 24, 2017)
8. GE to invest $1.4 billion in 3D printing. 2016. *Metal Powder Report*, 71:473.
9. Concept Laser develops 'modular' 3D printing. 2016. *Metal Powder Report*, 71:473.
10. Light cocoon features 3D printed parts. 2016. *Metal Powder Report*, 71:293.
11. Concept Laser 3D prints titanium beak for macaw. 2016. *Metal Powder Report*, 71:367.
12. Tantalum powder success for biomedical applications. 2016. *Metal Powder Report*, 71:207.
13. 3D printed titanium alloy could replace bone. 2016. *Metal Powder Report*, 71:288–289.
14. https://www.materialstoday.com/additive-manufacturing/news/slm-joint-venture-will-invest-in-aluminum/ (Accessed on October 24, 2017)
15. UK team unveils 3D printed mountain bike frame. 2016. *Metal Powder Report*, 71:361.

16. http://www.airbus.com/newsroom/press-releases/en/2016/05/APWorks-Launch-Light-Rider.html (Accessed on October 24, 2017)
17. New way to 3D print metals using PM ink. 2016. *Metal Powder Report*, 71:117–118.
18. http://www.thehindu.com/news/national/andhra-pradesh/nit-warangal-signs-mou-with-cmti-bangalore/article5840322.ece (Accessed on October 24, 2017)
19. UK creates first dedicated AM master's course. 2016. *Metal Powder Report*, 71:359.
20. https://3dprintingindustry.com/news/arizona-state-university-opening-additive-manufacturing-facility-concept-laser-102892/ (Accessed on October 24, 2017)
21. https://www.tctmagazine.com/3d-printing-news/eos-leading-business-school-emphasize-industrial-3d-printing/ (Accessed on October 24, 2017)
22. Slotwinski, J.A., Crane, E.E., Ramsburg, J.T., LaBarre, E.D., Forrest, R.J. 2016. Additive manufacturing at the Johns Hopkins University applied physics laboratory. *Johns Hopkins APL Technical Digest*, 33:225–234.
23. Mazzoli, A. 2013. Selective laser sintering in biomedical engineering. *Medical & Biological Engineering & Computing*, 51:245–256.
24. Manriquez-Frayre, I.A., Bourell, D.L. 1990. Selective laser sintering of binary metallic powder. *Proceedings of Solid Freeform Fabrication Symposium*, 99–106.
25. Olakanmi, E.O. 2013. Selective laser sintering/melting (SLS/SLM) of pure Al, Al–Mg, and Al–Si powders: Effect of processing conditions and powder properties. *Journal of Materials Processing Technology*, 213:1387–1405.
26. Gusarov, A.V., Laoui, T., Froyen, L., Titov, V.I. 2003. Contact thermal conductivity of a powder bed in selective laser sintering. *International Journal of Heat and Mass Transfer*, 46:1103–1109.
27. Paul, C.P., Bhargava, P., Kumar, A., Pathak, A.K., Kukreja, L.M. 2012. Laser rapid manufacturing: Technology, applications, modelling and future prospects. In: J. P Davim (editor-in-chief), *Lasers in Manufacturing*, Boston, MA: Wiley, pp. 1–60.
28. Kruth, J.P., Levy, G., Klocke, F., Childs, T.H.C. 2007. Consolidation phenomena in laser and powder-bed based layered manufacturing. *Annals of the CIRP*, 56:730–759.
29. Gu, D., Shen, Y. 2008. Processing conditions and microstructural features of porous 316L stainless steel components by DMLS. *Applied Surface Science*, 255:1880–1887.
30. Sames, W.J., List, F.A., Pannala, S., Dehoff, R.R., Babu, S.S. 2016. The metallurgy and processing science of metal additive manufacturing. *International Materials Reviews*, 61:1–46.
31. http://www.rpminnovations.com/index.php?page=news-article&id=5 (Accessed on October 24, 2017)
32. Syed, W.U.H., Li, L. 2005. Effects of wire feeding direction and location in multiplelayer diode laser direct metal deposition. *Applied Surface Science*, 248:518–524.
33. https://www.optomec.com (Accessed on October 24, 2017)
34. http://dm3dtech.com (Accessed on October 24, 2017)
35. https://www.eos.info (Accessed on October 24, 2017)
36. https://slm-solutions.com/products/machines (Accessed on October 24, 2017)

37. http://www.conceptlaserinc.com/en/products.html (Accessed on October 24, 2017)
38. Toyserkani, E., Khajepour, A., Corbin, S.F. 2004. *Laser Cladding*, New York: CRC Press.
39. Paul, C.P., Ganesh, P., Mishra, S.K., Bhargava, P., Negi, J., Nath, A.K. 2007. Investigating laser rapid manufacturing for Inconel-625 components. *Optics & Laser Technology*, 39:800–805.
40. Guan, K., Wang, Z., Gao, M., Li, X., Zeng, X. 2013. Effects of processing parameters on tensile properties of selective laser melted 304 stainless steel. *Materials and Design*, 50:581–586.
41. Carter, L.N., Attallah, M.M., Reed, R.C. 2012. Laser powder bed fabrication of nickel-base superalloys: Influence of parameters; characterisation, quantification and mitigation of cracking. *12th International Symposium on Superalloys*, Champion, PA, 9–13 September 2012.
42. Hanzl, P., Zetek, M., Bakša, T., Kroupa, T. 2015. The influence of processing parameters on the mechanical properties of SLM parts. *Procedia Engineering*, 100:1405–1413.
43. Shifeng, W., Shuai, L., Qingsong, W., Yan, C., Sheng, Z., Yusheng, S. 2014. Effect of molten pool boundaries on the mechanical properties of selective laser melting parts. *Journal of Materials Processing Technology*, 214:2660–2667.
44. Santomaso, A., Lazzaro, P., Canu, P. 2003. Powder flowability and density ratios: The impact of granules packing. *Chemical Engineering Science*, 58:2857–2874.
45. Sames, W.J., Medina, F., Peter, W.H., Babu, S.S., Dehoff, R.R. 2014. Effect of process control and powder quality on inconel 718 produced using electron beam melting. *8th International Symposium on Superalloy 718 and Derivatives*, John Wiley & Sons, pp. 409–423.
46. Attar, H., Calin, M., Zhang, L.C., Scudino, S., Eckert, J. 2014. Manufacture by selective laser melting and mechanical behaviour of commercially pure titanium. *Materials Science & Engineering A*, 593:170–177.
47. Barbas, A., Bonnet, A.-S., Lipinski, P., Pesci, R., Dubois, G. 2012. *Journal of Mechanical Behavior of Biomedical Materials*, 9:34–44.
48. Attar, H., Prashanth, K.G., Chaubey, A.K., Calin, M., Zhang, L.C., Scudino, S., Eckert, J. 2015. Comparison of wear properties of commercially pure titanium prepared by selective laser melting and casting processes. *Materials Letters*, 142:38–41.
49. Wauthle, R., van der Stok, J., Amin, Y.S., Humbeeck, J.V., Kruth, J.-P., Zadpoor, A.A., Weinans, H., Mulier, M., Schrooten, J. 2015. Additively manufactured porous tantalum implants. *ActaBiomaterialia*, 14:217–225.
50. Thijs, L., Sistiaga, M.L.M., Wauthle, R., Xie, Q., Kruth, J.-P., Humbeeck, J.V. 2013. Strong morphological and crystallographic texture and resulting yield strength anisotropy in selective laser melted tantalum. *ActaMaterialia*, 61:4657–4668.
51. Khan, M., Dickens, P.M. 2010. Selective laser melting (SLM) of pure gold. *Gold Bulletin*, 43:114–121.
52. VamsiKrishna, B., Bose, S., Bandyopadhyay, A. 2007. Low stiffness porous Ti structures for load-bearing implants. *ActaBiomaterialia*, 3:997–1006.
53. Balla, V.K., Bodhak, S., Bose, S., Bandyopadhyay, A. 2010. Porous tantalum structures for bone implants: Fabrication, mechanical and in vitro biological properties. *ActaBiomaterialia*, 6:3349–3359.

54. Dinda, G.P., Song, L., Mazumder, J. 2008. Fabrication of Ti-6Al-4V scaffolds by direct metal deposition. *Metallurgical and Materials Transactions A*, 39a: 2914–2922.
55. Shah, K., Pinkerton, A.J., Salman, A., Li, L. 2010. Effects of melt pool variables and process parameters in laser direct metal deposition of aerospace alloys. *Materials and Manufacturing Processes*, 25:1372–1380.
56. Zhao, X., Li, S., Zhang, M., Liu, Y., Sercombe, T.B., Wang, S., Hao, Y., Yang, R., Murr, L.E. 2016. Comparison of the microstructures and mechanical properties of Ti–6Al–4V fabricated by selective laser melting and electron beam melting. *Materials and Design*, 95:21–31.
57. Khorasani, A.M., Gibson, I., Goldberg, M., Littlefair, G. 2016. A survey on mechanisms and critical parameters on solidification of selective laser melting during fabrication of Ti-6Al-4V prosthetic acetabular cup. *Materials and Design*, 103:348–355.
58. Jia, Q., Gu, D. 2014. Selective laser melting additive manufacturing of Inconel 718 superalloy parts: Densification, microstructure and properties. *Journal of Alloys and Compounds*, 585:713–721.
59. Amato, K.N., Gaytan, S.M., Murr, L.E., Martinez, E., Shindo, P.W., Hernandez, J., Collins, S., Medina, F. 2012. Microstructures and mechanical behavior of Inconel 718 fabricated by selective laser melting. *ActaMaterialia*, 60:2229–2239.
60. Tian, Y., Mcallister, D., Colijn, H., Mills, M., Farson, D., Nordin, M., Babu, S. 2014. Rationalization of microstructure heterogeneity in INCONEL 718 builds made by the direct laser additive manufacturing process. *Metallurgical and Materials Transactions A*, 45a:4470–4483.
61. Bartkowiak, K., Ullrich, S., Frick, T., Schmidt, M. 2011. New developments of laser processing aluminium alloys via additive manufacturing technique. *Physics Procedia*, 12:393–401.
62. Buchbinder, D., Schleifenbaum, H., Heidrich, S., Meiners, W., Bültmann, J. 2011. High power selective laser melting (HP SLM) of aluminum parts. *Physics Procedia*, 12:271–278.
63. Brandl, E., Heckenberger, U., Holzinger, V., Buchbinder, D. 2012. Additive manufactured AlSi10Mg samples using selective laser melting (SLM): Microstructure, high cycle fatigue, and fracture behaviour. *Materials and Design*, 34:159–169.
64. Li, Y., Gu, D. 2014. Parametric analysis of thermal behavior during selective laser melting additive manufacturing of aluminum alloy powder. *Materials and Design*, 63:856–867.
65. Thijs, L., Kempen, K., Kruth, J.-P., Humbeeck, J.V. 2013. Fine-structured aluminium products with controllable texture by selective laser melting of pre-alloyed AlSi10Mg powder. *ActaMaterialia*, 61:1809–1819.
66. Sing, S.L., Yeong, W.Y., Wiria, F.E. 2016. Selective laser melting of titanium alloy with 50 wt% tantalum: Microstructure and mechanical properties. *Journal of Alloys and Compounds*, 660:461–470.
67. Tolosa, I., Garciandía, F., Zubiri, F., Zapirain, F., Esnaola, A. 2010. Study of mechanical properties of AISI 316 stainless steel processed by 'selective laser melting', following different manufacturing strategies. *International Journal of Advanced Manufacturing Technology*, 51:639–647.
68. Bayode, A., Akinlabi, E.T., Pityana, S. 2016. Characterization of laser metal deposited 316L stainless steel. *Proceedings of the World Congress on Engineering*. London, UK.

69. Mazumder, J., Choi, J., Nagarathnam, K., Koch, J., Hetzner, D. 1997. The direct metal deposition of H13 tool steel for 3-D components. *Journal of the Minerals, Metals, and Materials Society*, 49:55–60.
70. Pinkerton, A.J., Li, L. 2005. Direct additive laser manufacturing using gas- and water-atomised H13 tool steel powders. *International Journal of Advanced Manufacturing Technology*, 25:471–479.
71. Kac, S., Kusiński, J. 2003. SEM and TEM microstructural investigation of high-speed tool steel after laser melting. *Materials Chemistry and Physics*, 81:510–512.
72. Kempen, K., Thijs, L., Vrancken, B., Buls, S., Humbeeck, J.V., Kruth, J.P. 2013. Producing crack-free, high density M2 Hss parts by selective laser melting: Pre-heating the baseplate. *Proceedings of Solid Freeform Fabrication Symposium*, pp. 131–139. University of Texas, Austin, TX: Annual International Solid Freeform Fabrication Symposium.
73. Kobryn, P.A., Moore, E.H., Semiatin, S.L. 2000. The effect of laser power and traverse speed on microstructure, porosity, and build height in laser-deposited Ti-6Al-4V. *Scriptamaterilia*, 43:299–305.
74. Read, N., Wang, W., Essa, K., Attallah, M.M. 2015. Selective laser melting of AlSi10Mg alloy: Process optimisation and mechanical properties development. *Materials and Design*, 65:417–424.
75. Scientists discover 'big step forward' for metal 3D printing. 2016. *Metal Powder Report*, 71:365–366.
76. Gu, D., Shen, Y. 2009. Balling phenomena in direct laser sintering of stainless steel powder: Metallurgical mechanisms and control methods. *Materials and Design*, 30:2903–2910.
77. Li, R., Liu, J., Shi, Y., Wang, L., Jiang, W. 2012. Balling behavior of stainless steel and nickel powder during selective laser melting process. *International Journal of Advanced Manufacturing Technology*, 59:1025–1035.
78. Yasa, E., Poyraz, O., Solakoglu, E.U., Akbulut, G., Oren, S. 2016. A study on the stair stepping effect in direct metal laser sintering of a nickel based super alloy. *Procedia CIRP*, 45:175–178.
79. Mingareev, I., Bonhoff, T., El-Sherif, A.F., Meiners, W., Kelbassa, I., Biermann, T., Richardson, M. 2013. Femtosecond laser post-processing of metal parts produced by laser additive manufacturing. *Journal of Laser Applications*, 25:1–4.
80. Gibson, I., Rosen, D.W., Stucker, B. 2010. *Additive Manufacturing Technologies* (Vol. 238). New York: Springer.
81. Jinoop, A.N. 2016. Master's Thesis, Investigation on laser shock processing of direct metal laser sintered inconel 718, National Institute of Technology Warangal, India.
82. Shiva, S. 2017. Ph.D. Thesis, Laser additive manufacturing of bulk shape memory alloy structures: Numerical modeling and experimental investigation. Indian Institute of Technology Indore, India.
83. Chlebus, E., Gruber, K., Kuźnicka, B., Kurzac, J., Kurzynowski, T. 2015. Effect of heat treatment on the microstructure and mechanical properties of Inconel 718 processed by selective laser melting. *Materials Science & Engineering A*, 639:647–655.
84. Thöne, M., Leuders, S., Riemer, A., Tröster, T., Richard, H.A. 2012. Influence of heat-treatment on selective laser melting products–e.g. Ti6Al4V. *Proceedings of Solid Freeform Fabrication Symposium*, pp. 492–498. University of Texas, Austin, TX: Annual International Solid Freeform Fabrication Symposium.

85. Zhang, D., Niu, W., Cao, X., Liu, Z. 2015. Effect of standard heat treatment on the microstructure and mechanical properties of selective laser melting manufactured Inconel 718 superalloy. *Materials Science & Engineering A*, 644:32–40.

86. Zhao, X., Lin, X., Chen, J., Xue, L., Huang, W. 2009. The effect of hot isostatic pressing on crack healing, microstructure, mechanical properties of Rene88DT superalloy prepared by laser solid forming. *Materials Science and Engineering A*, 504:129–134.

87. Manfredi, D., Calignano, F., Ambrosio, E.P., Krishnan, M., Canali, R., Biamino, S., Pavese, M. et al. 2013. Direct Metal Laser Sintering: an additive manufacturing technology ready to produce lightweight structural parts for robotic applications. *La MetallurgiaItaliana*, 10:15–24.

88. Bourell, D.L., Vallabhajosyula, P., Stevinson, B., Chen, S., Beaman, J.J. Jr. 2008. Rapid manufacturing using infiltration selective laser sintering. *Proceedings of the 9th Biennial Conference on Engineering Systems Design and Analysis*, pp. 173–178. Israel: ASME.

89. Sindel, M., Pintat, T., Greul, M., Nyrhila, O., Wilkening, C. 1994. Direct selective laser sintering of metals and metal melt infiltration for near net shape fabrication of components. *Proceedings of Solid Freeform Fabrication Symposium*, pp. 94–101. University of Texas, Austin, TX: Annual International Solid Freeform Fabrication Symposium.

90. Kruth, J.-P., Vandenbroucke, B., Vaerenbergh, J.V., Mercelis, P. 2005. Benchmarking of different SLS/SLM processes as rapid manufacturing techniques. *International Conference Polymers & Moulds Innovations*, pp. 1–6. Belgium: International Conference Polymers and Moulds Innovations.

91. Yasa, E., Demir, F., Akbulut, G., Cızıoğlu, N., Pilatin, S. 2014. Benchmarking of different powder-bed metal fusion processes for machine selection in additive manufacturing. *Proceedings of Solid Freeform Fabrication Symposium*, pp. 390–403. University of Texas, Austin, TX: Annual International Solid Freeform Fabrication Symposium.

92. Cooke, A.L., Soons, J.A. 2010. Variability in the geometric accuracy of additively manufactured test parts. *Proceedings of Solid Freeform Fabrication Symposium*, Austin, TX, pp. 1–12.

93. Vandenbroucke, B., Kruth, J.-P. 2007. Selective laser melting of biocompatible metals for rapid manufacturing of medical parts. *Rapid Prototyping Journal*, 13:196–203.

94. Shiva, S., Palani, I.A., Mishra, S.K., Paul, C.P., Kukreja, L.M. 2015. Investigations on the influence of composition in the development of Ni–Ti shape memory alloy using laser based additive manufacturing. *Optics & Laser Technology*, 69:44–51.

95. Shiva, S., Palani, I.A., Paul, C.P., Mishra, S.K., Singh, B. 2016. Investigations on phase transformation and mechanical characteristics of laser additive manufactured TiNiCu shape memory alloy structures. *Journal of Materials Processing Technology*, 238:142–151.

96. Shiva, S., Palani, I.A., Paul, C.P., Singh, B. 2016. Laser annealing of laser additive–manufactured Ni–Ti structures: An experimental–numerical investigation. *Proceedings of the Institution of Mechanical Engineers, Part B: Journal of Engineering Manufacture*, pp. 1–14. Thousand Oaks, CA: Sage Publishers.

chapter three

Investigations to enhance the strength of open cell porous regular interconnect structure

Jatender Pal Singh and Pulak M. Pandey

Contents

3.1   Introduction

Selective Laser Sintering (SLS) is a powder-based Additive Manufacturing (AM) process used for fabrication of functional parts. The fabrication process involves sintering of powder of one layer over the other with the use of laser. The powders of polymer, ceramic, composites nylon-glass, polymer-metal, cermets, metals, alloys, steels and wax are being used in this process [1]. The fabrication process utilizes the STL file format, which is sliced in the layers of thickness ranging from 0.05 to 0.3 mm, depending upon the specifications of the machine. Support structure is not being used, as unsintered powder acts as support to the fabricated part. The advantages of the SLS process are appended as follows:

- Requires no support structure
- Provides wide range of material that can be processed
- Allows ability to fabricate functional parts in small batches
- Requires little post processing
- Does not require any post curing
- Is useful in rapid tooling applications

3.2   Fabrication process

The fabrication process starts with the modelling of the CAD structure. The part file is tessellated to the STL format. This format may contain defects like bad edges, intersecting triangles, flipped triangles, planar hole, etc., which can be removed using software like RP MAGICS™. Thereafter the file is sliced into a number of layers (virtual). The sliced part is loaded to the machine. The operation of the machine starts with the movement of a recoater, which is used to spread the powder on the bed. The bed is enclosed in a chamber wherein the temperature is maintained below the melting point of the powder. Infrared heaters are used in the chamber to maintain the temperature. Thereafter the laser is exposed to the specific sliced area in accordance with the CAD model, and melting of the powder takes place. The platform descends and the recoater again spreads the powder, and the process repeats till completion of part [2]. The post-processing of the part is also done to some extent, which includes cleaning of the powder with the help of compressed air. The steps of the fabrication process are shown in Figure 3.1.

The process retains heat from two sources – first, from Infra red heaters, which are used to maintain powder bed temperature and second, the source of heat energy is delivered by a scanning laser, which completes the sintering process at selected regions. This combination helps to minimize thermal gradients and facilitates fusion of a sintered layer with the previous. The densification/consolidation or fusion of layer is dependent

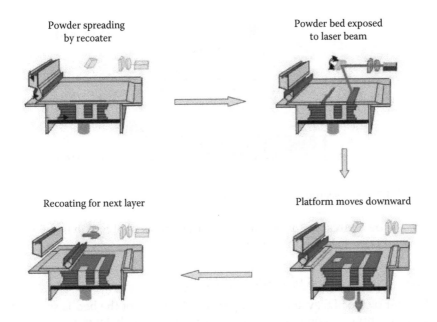

Figure 3.1 Typical SLS process cycle. (From EOSINT P380, *Basic Training Manual*, EOS Gmbh, Munich, Germany, 2003.)

upon the properties of the powder, the part bed temperature and process parameters. There should be minimum difference of temperature between the sintered polymer and the new powder layer. Curling of layer takes place when the powder is too cold while it flows and spreads when powder is too hot. In order to have minimum curling, distortion and shrinkage, the heat input as well as other process parameters must be optimised. The energy density, which is also called Andrew number, is an important parameter of the SLS process and is represented by the equation:

$$E = \frac{P}{V \times H_s} \tag{3.1}$$

where:

E is the energy density J/mm^2
P is the laser power in watt
V is the beam speed in mm/s
H$_s$ is the hatch spacing in mm

It is a combination of laser power, beam speed and hatch spacing. A specific range of values for energy density is required to sinter the powder. There are many issues associated with laser scanning strategies, which

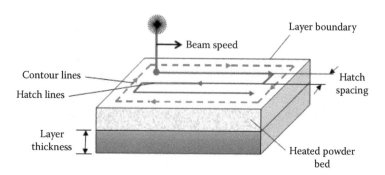

Figure 3.2 Laser scanning process in SLS.

need to be addressed for obtaining accurate parts in SLS. The laser scanning process is shown in Figure 3.2.

3.3 Kinematics of laser scanning

In most of the SLS systems, the uppermost surface of the bed is scanned with hatching in combination with countering. The hatching is done in raster scan mode to cover the area of cross section while the contouring is performed by scanning boundary outline in vector scan mode. The layout of scanning in hatching mode is dependent upon the geometry of the cross section and orientation [2]. Figure 3.3 depicts the same.

3.3.1 Scanning path

A variety of scanning strategy is available in the SLS process. It basically includes scan patterns of the laser in linear x or y or the alternation of both. There are four types of common exposures available in the SLS system:

1. Parallel to X direction
2. Parallel to Y direction

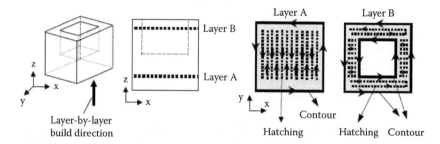

Figure 3.3 Exposure strategy showing hatching and contouring.

3. Along both X and Y direction
4. Along X direction for the first layer, subsequently along Y direction for the next layer

3.3.2 Beam compensation

Laser beam has small diameter and is referred as laser diameter. When laser strikes at a point, its influence extends beyond laser diameter. This region of influence is usually called a spot diameter. The region of influence depends upon the laser energy density used. The boundary of the part is fabricated using contour lines while for inside fabrication, hatch lines are used. The contour lines use low laser power and high beam speed for scanning in comparison to the hatch lines. It is recommended to offset laser beam for compensation of spot diameter. The beam offset values for contour and hatch lines are in variance. In the SLS system, beam offset can be chosen for contouring and hatching separately. In order to expose the edge of the boundary during contouring, the value of the beam offset, (d_c) is limited to half of the contour spot diameter. The same is shown in Figure 3.4.

In case the beam offset is adjusted to less or greater than half of the effective beam diameter in coutour, there is the possibility of not sintering the part at the intended region of sintering outside the layer edge. It would deviate the dimensional accuracy of the part. During hatching, the initial beam offset value is changed again with respect to the edge of the boundary (which should be larger than that for contouring). However, caution must be observed that there should not be any unsintered particle

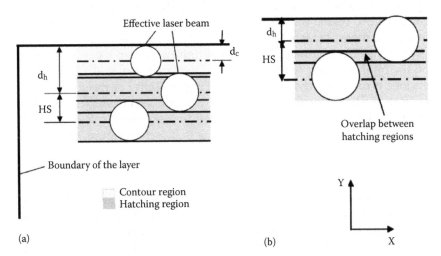

Figure 3.4 Exposure strategies for beam compensation: (a) contouring and hatching and (b) hatching.

left between the contour path and the hatching region. Thus, the beam offset for hatching (d_h) must be chosen in such a way so as to form narrow overlapping regions between the contour path and hatching region. The overlap should not be too wide so that over-sintering can be prevented. When hatch lines are exposed, the value of beam offset for hatching (d_h) is chosen as half of the spot diameter.

3.3.3   *Inertia of scanning mirror*

The laser galvanometric scanning technique is used widely in SLS due to its high speed and flexibility. A galvanometric scanner in combination with a mirror, servo-actuated limited rotation motor are the major components in a galvanometric scanning system. It deflects a beam to follow a defined path in a controlled fashion. On entrance of a beam into the scanning system, it encounters the X scanner/mirror followed by the Y scanner/mirror. It undergoes deflection and finally focuses in the field of view. In the SLS process, every layer of scan path is converted into a series of vectors using a control subsystem and scanning routine. The vector sets are stored in the control subsystem and are transferred into a series of control signals to drive the X and Y scanners. Usually there are variations in scanning speed due to acceleration and deceleration of the scanning mirror during hatching exposure at the boundaries of the layer, which are to be eliminated to get a uniform energy density. In order to avoid inconsistencies in the energy density of exposure, the length of scan should be compensated during acceleration and deceleration of galvano mirrors. The laser should be switched off while scanning these lengths so that no exposure is to be made on compensated length. This length of unexposed compensated region is independent of part size, shape and location. Some of the commercial machines usually call this kind of exposure strategy as skywriting. The accuracy of the laser scanning system has a direct effect on the final accuracy of the part to be fabricated. Xie et al. [3] discussed various errors involved in the positioning of ability of the scanner with respect to RP/RM processes. It is very important to identify all kinds of errors and their effects before the scanning accuracy can be correctly interpreted. The schematic of skywriting is shown in Figure 3.5.

3.3.4   *Positioning ability of the scanner*

Process software is used to plan laser path. Certain errors are inculcated into the shrinkage pattern due to the low positioning ability of the scanner hardware. In the SLS process, dedicated computer software is used to scan the laser path computed from a layered file. The approximations of hatch line generation algorithms cause dimensional errors and alter the shrinkage pattern of the specimens. It requires correction. The exposed region is superimposed

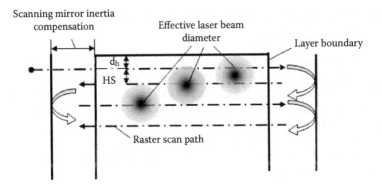

Figure 3.5 Skywriting exposure strategy.

over a fixed two-dimensional grid pattern. Then the first and last scan lines are aligned to pass through the grid lines rather than the actual occupied space by the material due to shrinkage effect and the original intended dexel space. This approach is mainly driven by the inability of the scanner to position the beam to the intended dexel space thereby depending upon the positioning ability of the galvano scanner used in the SLS machine. The first and last hatch lines are snapped to this grid lines, and hatch lines are generated in between based on the hatch spacing value input to the machine. The nominal dimension fabricated can be bigger or smaller than the original CAD dimension depending on the length of the specimen. Figure 3.6 shows the same.

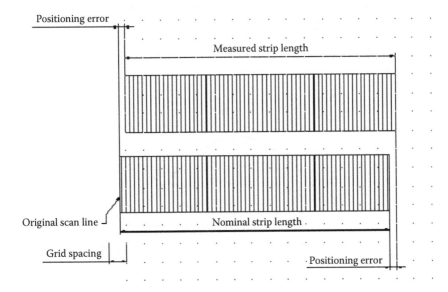

Figure 3.6 Positioning error in hatching.

3.4 Quantitative analysis of laser in selective laser sintering

The selective laser sintering process uses laser power to control the solidification and accuracy of the fabricated prototype. The parts are built layer by layer in SLS. Therefore, the researchers have much interest in calculating energy of laser at depth z [1]. Laser has a capability to produce polymerization by irradiance in comparison to the regular arc lamp. The energy of the laser does not remain constant as it travels in the powder but decays exponentially in accordance with the Beer-Lambert law of absorption enumerated as follows:

$$H_{(x,y,z)} = H_{(x,y,0)} \exp\left(-\frac{z}{D_p}\right) \tag{3.2}$$

where:

$H_{(x,y,z)}$ is the exposure at position x, y, z
$H_{(x,y,0)}$ is the exposure at bottom x, y, 0
z is the any height
D_p is the powder constant defined by the depth of particular powder
     that results in reduction of irradiance level to 1/e (=1/2.718) of
     the H_0 level on the surface

It means that at a depth $z = D_p$, the irradiance is $1/2.718 = 37\%$ of H_0. The detail of laser energy distribution is shown in Figure 3.7. W_0 is the radius of central spot size in mm.

The Gaussian curve controls the physics of laser energy distribution. The decay of the laser energy across the surface is given by the equation:

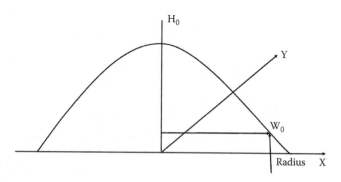

Figure 3.7 Gaussian decay of laser energy across the surface.

$$H_{(x,y,0)} = H_{(r,0)} = H_0 \, \exp\left(-\frac{2r^2}{W_0^2}\right) \tag{3.3}$$

where $W_0 = 1/e^2$. The Gaussian half-width $r = W_0$,

$$H = H_0 e^{-2} = 0.135 \, H_0 \tag{3.4}$$

If P_L is the nominal laser power, $H_{average}$ can be given as

$$H_{average} = \frac{P_L}{\pi W_0^2} \tag{3.5}$$

If scanning speed (V) is given the laser exposure time (t_e) on a given area is given by

$$t_e = \frac{2W_0}{V} \tag{3.6}$$

The average energy density ($E_{average}$) of laser exposure is given by

$$E_{average} = H_{average} \times t_e \tag{3.7}$$

In order to calculate the polymerization ability, Plank's equation may be used given as under:

$$E_{(photon)} = \frac{hc}{\Lambda} \tag{3.8}$$

where:
   E is the photon energy
   h is the Plank's constant
   c is the velocity of light
   Λ is the laser wavelength

Further, the number of photons per unit area hitting the resin can be calculated using the equation:

$$N_{ph} = \frac{E_{Average}}{E_{photon}} \tag{3.9}$$

3.5   Strength related studies for solid structure

Various attempts were carried out for parametric study of the SLS process on mechanical properties. Nelson et al. [4] developed a heat transfer model in one dimension for the SLS process. Prediction on sintering depths of

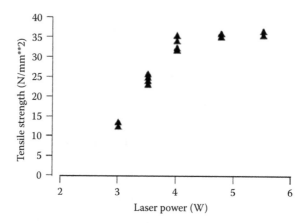

Figure 3.8 Variation of tensile strength with laser power. (From Gibson, I. and Shi, D., *Rapid Prototyping J.*, 3, 129–136, 1997.)

bisphenol polycarbonate powders was carried out. Experimentations were also carried out to validate the findings of the simulation studies. They showed a qualitative correlation between the energy density and the observed degree of sintering. They proposed a dimensional group in which the energy density was based on the major process parameters – namely, laser power, beam speed and scan spacing. Gibson and Shi [5] experimented with fine nylon material to investigate the effect of SLS parameters such as hatch spacing, laser power and laser beam speed on the density and strength of the fabricated parts. It was exhibited that density and strength of the SLS part increase up to a particular level (Figure 3.8) with the increase in laser power. Thereafter, it decreases with the increase in scan spacing. Build orientation and scan length were found to be as effective parameters that affect the part strength. Improved mechanical properties of the SLS parts were observed, which were fabricated with short scan vector. It resulted in uniform temperature distribution.

Thompson and Crawford [6] investigated the effect of SLS parameters – namely, build orientation, layer thickness and laser power on the tensile strength and surface roughness of the part using polycarbonate. Regression models were developed to predict surface roughness and tensile strength. Figure 3.9 shows response surfaces reported in their work. The results showed an increase in part strength with an increase in laser power, while a decrease in strength was reported with an increase in layer thickness.

Williams and Deckard [7] proposed a numerical model to simulate process phenomena and also studied the effect of process parameters on part strength and density. It was reported that delay time and spot size had a great effect on the density as well as on the part strength. An increase in

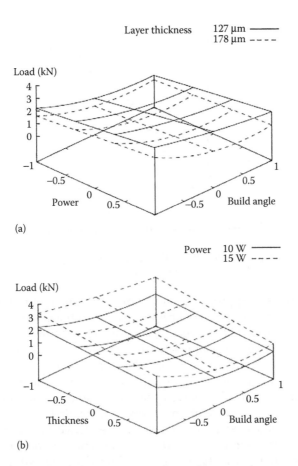

Layer thickness 127 µm ——
178 µm ----

(a)

Power 10 W ——
15 W ----

(b)

Figure 3.9 Tensile strength response surfaces: (a) strength versus power and build angle and (b) strength versus layer thickness and build angle. (From Thompson, D.C. and Crawford, R.H., *J. Manuf. Syst.*, 16, 273–289, 1997.)

delay period causes a decrease in average density and strength but with the increase in laser spot size, part strength and density were found to be increased. The existence of an optimum delay range was observed, which had resulted from SLS process parameters, namely beam speed, laser power and hatch spacing. This delay range contributed to have maximum SLS part strength in terms of flexural modulus for bisphenol polycarbonate material. Miller et al. [8] discussed a concept of an energy delivery system for the polymer coated steel powder in the SLS process. They conducted experiments using two level-fixed factorial designs of experiments to predict part strength as a function of processing parameters and developed a parametric equation for this purpose. They concluded that

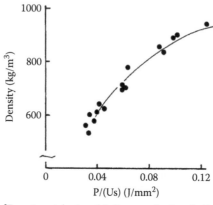

[Experimental values (•), Computed values (—)]

Figure 3.10 Variation of density with respect to energy density. (From Childs, T.H.C. et al., *J. Eng. Manuf. B*, 213, 333–349, 1999.)

an energy density approach to model SLS did not account for the delay time and reported the effect of heat loss by convection and radiation. They proposed a new energy delivery system, which consisted of a large laser spot size to fill open areas quickly and a small laser spot size for outline and fine details. Childs et al. [9] had done analysis on powder densification and thermal modelling of amorphous polycarbonate in an SLS process. The analysis results showed that linear accuracy and densification were found to be dependent on energy density. Bed density of powder and layer thickness were also considered effective for linear accuracy. Simulation was done using fixed finite element methods and adaptive mesh finite differences. The experimental results validated the predicted density of the part. The same is presented in Figure 3.10.

Tontowi and Childs [10] estimated the densities of Duraform (nylon-12) and Protoform (glass filled nylon-11), which were available commercially. The densities of sintered SLS parts were studied by varying part bed temperature. A density prediction model was also developed based on the experimental results (Figure 3.11).

Chatterjee et al. [11] experimented with metallic powders wherein responses like density, porosity and hardness were studied by varying hatch spacing and layer thickness. An increase in porosity was observed due to an increase in layer thickness and hatch spacing. Figure 3.12 shows the same.

Dewidar et al. [12] investigated processing conditions of SLS for pre-alloyed high-speed steel powder. They fabricated single-layered and multiple-layered parts by adopting different scanning strategies and conducted a four-point bend test to measure mechanical properties and density. They concluded that SLS of high-speed steel alloyed to bronze infiltration could produce material for load-bearing applications. The infiltration

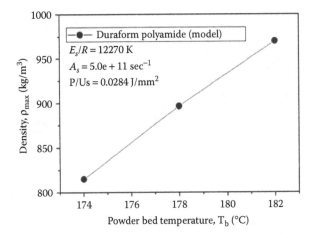

Figure 3.11 Density versus powder bed temperature. (From Tontowi, A.E. and Childs, T.H.C., *Rapid Prototyping J.,* 7, 180–184, 2001.)

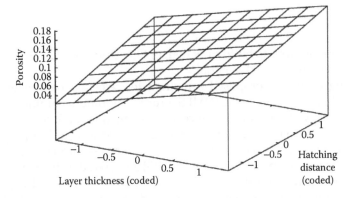

Figure 3.12 Variation in porosity with layer thickness and hatching distance. (From Chatterjee, A.N. et al., *J. Mater. Process. Technol.,* 136, 151–157, 2003.)

facilitated the parts to attain modulus of 92 GPa and bending strength of 460 MPa. The mechanical properties were found comparable with aluminium alloys. The scope of enhancing the mechanical properties of the SLS part by infiltration was also confirmed. With the processing conditions, HSS powder could not be processed to the stage where it could be used for load-bearing applications. Storch et al. [13] studied the mechanical properties of two commercial materials, namely EOS Direct steel 20 and 3D LaserForm ST100 in comparison to conventional materials – that is, aluminium alloys used in automotive applications. Their findings have been presented in Figure 3.13. They observed that material properties

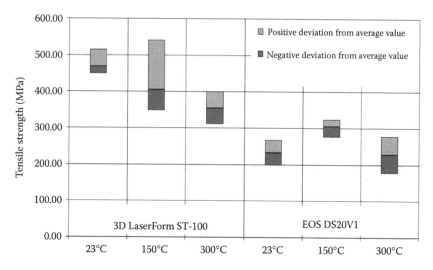

Figure 3.13 Variation of tensile strength with temperature. (From Storch, S. et al., *Rapid Prototyping J.*, 9, 240–251, 2003.)

were more sensitive to build direction for EOS Direct Steel 20 as compared to 3D Laser Form ST100 material.

Ning et al. [14] proposed a method for selecting parameters for a Direct Metal Laser Sintering (DMLS) process. Optimum parameters could be chosen by the users to reduce processing time and surface roughness, as well as enhance geometric accuracy and mechanical properties. A model was developed to achieve part properties through better mapping of process parameters. Ning et al. [15] analysed the effect of variation in hatch length on heterogeneity and anisotropy using a DMLS process. It was found that heterogeneity observed in the part was due to short hatch lines. Ning et al. also concluded that orientation and hatch direction affected the anisotropy. Orientation showed greater influence on the part strength as compared to hatch direction. Figure 3.14 shows various orientations of tensile specimens built with loading direction and build direction. Corresponding strength values for each specimen are shown in Figure 3.15. It was concluded that the part built with short hatch lines had higher strength than that built with the long hatch lines.

Zarringhalam et al. [16] experimented with fresh and refreshed powder to fabricate SLS parts. Thermal, molecular and microstructure analysis were done to ascertain effects of process parameters on part strength. No effect of refresh rate on part strength was reported. Ajoku et al. [17] experimented and developed a finite element model for compressive strength. Compressive strength of SLS parts against injection-moulded

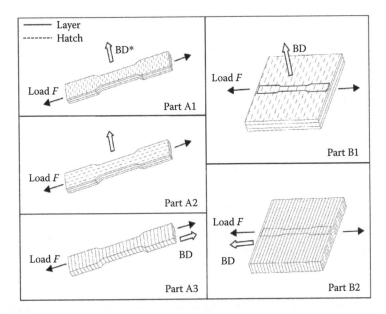

Figure 3.14 Specimens fabricated with different build directions. *BD, build direction. (From Ning, Y. et al., *J. Eng. Manuf.*, 219, 15–25, 2005.)

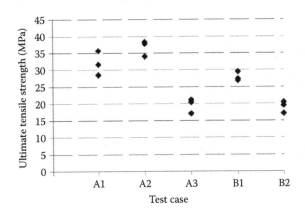

Figure 3.15 Variation in ultimate tensile strength for different build directions. (From Ning, Y. et al., *J. Eng. Manuf.*, 219, 15–25, 2005.)

parts of nylon-12 were compared. Ten percent less elastic modulus was observed in laser-sintered parts. High porosity arising due to non-uniform heat distribution was the main cause of low modulus in laser-sintered parts. Caulfield et al. [18] studied the effect of energy density on mechanical properties of polyamide SLS parts. The parts fabricated with high

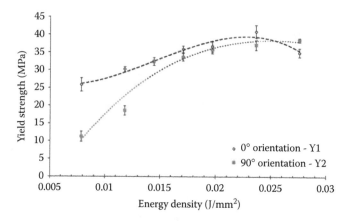

Figure 3.16 Yield strength values at different energy densities. (From Caulfield, B. et al., *J. Mater. Process. Technol.*, 182, 477–488, 2007.)

energy density showed more ductile behaviour than low energy density. Young's modulus, fracture strength and yield strength were also reported to be on the higher side with the increase in energy density. Figure 3.16 shows yield strength variation with respect to energy density.

3.6   Strength of foam

The foam can be categorized as open cell foam or closed cell foam. The foam that has pores visible from the outside and that may have interconnectivity or otherwise is called open cell foam. While the foam in which the pores are not visible from the outside despite their interconnectivity is termed as close cell foam. The study of properties in this chapter is restricted to the open cell foam only.

3.6.1   Open cell foam

Gibson and Ashby [19] experimented extensively to drive the relations of polymeric fabricated from most of the conventional processes such as bubbling of gas, stirring of foaming agent, dissolution of gas, etc., which fits to the experimental data. The mechanical properties are defined with the subscripted 's' – this means property of solid material of which the foam is made. For example, ρ_s is the density of solid material. The scaling relationships are given in Table 3.1.

Table 3.1 Scaling laws for mechanical properties of open cell foam

S. No.	Mechanical properties	Scaling relation
1	Young's modulus (GPa), E	$E = (0.1 - 4) E_s \left(\dfrac{\rho}{\rho_s} \right)^2$
2	Shear modulus (GPa), G	$G \approx \dfrac{3}{8} E$
3	Bulk modulus (GPa), K	$K \approx 1.1E$
4	Flexural modulus (GPa), E_f	$E_f \approx E$
5	Poisson ratio, υ	$0.32 - 0.34$
6	Compressive strength (MPa), σ_c	$\sigma_c = (0.1 - 1.0) \sigma_{c,s} \left(\dfrac{\rho}{\rho_s} \right)^{3/2}$
7	Tensile strength (MPa), σ_t	$\sigma_t \approx (1.1 - 1.4) \sigma_c$
8	Endurance limit (MPa), σ_e	$\sigma_e \approx (0.5 - 0.75) \sigma_c$
9	Densification strain, ε_D	$\varepsilon_D = (0.9 - 1.0) \times \left[1 - 1.4 \dfrac{\rho}{\rho_s} + 0.4 \left(\dfrac{\rho}{\rho_s} \right)^3 \right]$
10	Loss coefficient, η	$\eta = (0.95 - 1.05) \times \dfrac{\eta_s}{(\rho/\rho_s)}$
11	Hardness (MPa), H	$H = \sigma_c \left(1 + 2 \dfrac{\rho}{\rho_s} \right)$

3.6.2   Open cell porous regular interconnected structure

An Open Cell Porous Regular Interconnected Structure (OCPRIS) is open cell foam with all interconnected pores with each other at an equal interval of distance. All the pores are assessable from outside. This kind of structure can be fabricated through wax template and Rapid Prototyping (RP) processes. Therefore, the same structure can easily be fabricated through an SLS process.

It is evident from the study of solid structure fabricated through SLS that laser power, scan speed, hatch spacing and layer thickness play an important role in enhancement of strength. In order to investigate their contribution to the strength of OCPRIS, planning and experimentations were conducted by Singh et al. [21]. An EOSINT P380 (Germany) SLS machine was used with PA-2200 (purchased from EOSINT, Germany) having a size of 50–60 μm to fabricate the specimens. This SLS machine has a spot diameter of 0.6 mm. The feasibility study for pore fabrication was carried out so that clogged pores could be avoided. Thereafter specimens of designed porosity and dimensions in accordance with ASTM695D [20] were modelled and fabricated by varying process parameters, namely laser power, scan speed, layer thickness and hatch spacing. Response Surface Methodology with central composite design was used to map the experiments. A compression test was performed on a universal testing machine (INSTRON 5582, U.S.A). A regression equation was developed, and Analysis of Variance (ANOVA) was conducted to find contribution of each parameter on porous part strength. A trust region algorithm was used to optimize process variables. A few specimens were fabricated with optimized parameters, and a compression test was performed again to validate the results.

3.6.2.1 *Feasibility study of unclogged pore fabrication*

In order to explore fabrication of unclogged pores using an SLS process, computer-aided design (CAD) models of three-dimensional interconnected open porous mesh structures with square pore size of 0.2 × 0.2, 0.5 × 0.5, 0.8 × 0.8, 1 × 1 and 1.2 × 1.2 mm^2 were modelled using Solid Works 2004. Physical models as per the CAD design were fabricated using polyamide (PA-2200) by using an amalgamation of various process parameters. The pore sizes of 0.2 × 0.2, 0.5 × 0.5 and 0.8 × 0.8 mm^2 could not be obtained due to undue scuffing of partly sintered/unsintered powder. Further cleaning of holes could not be performed by any means. The cleaning of pores with size of 1×1 mm^2 was also found to be difficult as the holes were moderately blocked due to sintering. However, pores with size 1.2 ×1.2 mm^2 were easily obtained on cleaning with compressed air. The pore size 1.2 ×1.2 mm^2 was selected for fabrication in accordance with the requirement of ASTM695D shown in Figure 3.17,

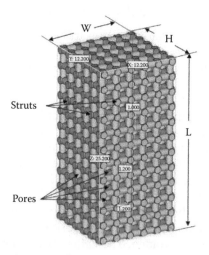

Figure 3.17 CAD model of 3-D mesh.

Table 3.2 Details of specimen

S. No.	Characteristics	Units	Value
1	L	mm	25.2
2	W and H	mm	12.2
3	Pore size	mm²	1.2 × 1.2
4	Strut	mm	1
5	Volume	mm³	1500
6	Surface area	mm²	4985
7	Porosity	Percent (%)	60

and details of dimensions are presented in Table 3.2. Thereafter the part was tessellated and saved in STL (stereo lithographic) format. Using Magics v 9.54 software (Materialise, Belgium), the errors of the STL file were edited and removed.

The orientation of CAD models during fabrication using SLS are shown in Figure 3.18. The parts were oriented in such a way to attain minimum Z-height so that minimum build time could be arrived at. The slicing was done virtually to facilitate layer-by-layer deposition of part.

The sliced parts were fabricated with a refresh rate of 30:70 (fresh and used powder). The part bed temperature was set at 176°C. Ranges of laser power (P), scan speed (V), hatch spacing (H_s) and layer thickness (L_t) were well thought-out to achieve the required level of energy density. It has been experienced that sintering does not occur if the value of energy density is lower than 1 J/cm² and degradation of polymer starts above 6 J/cm². Based on pilot experimentation and literature review, the values and ranges of hatch spacing, laser power, scan speed and layer thickness were determined and are presented in Table 3.3 for which minimum and maximum values of the energy density vary from 1.45 to 5.8 J/cm².

The laser power in an SLS EOSINT P380 machine is rated in percentage of available power and can be changed to W using the equation $y = -0.013x^2 + 1.559x - 0.430$ with correlation coefficient $R^2 = 0.999$ (Figure 3.19).

With the use of RSM and a central composite design, 31 experiments were planned. The levels and sequence to carry out experiments were obtained using MINITAB 13. The randomness in sequence of experiments and homogeneity was ensured. Five specimens for each designed experiment were fabricated using EOSINT P380, SLS machine. A total number of 155 specimens were obtained. Figure 3.20 shows the fabricated specimens.

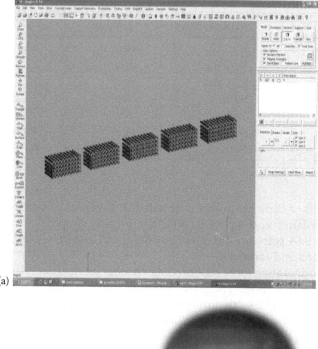

Figure 3.18 Fabrication of specimens using SLS: (a) orientation of the CAD specimens and (b) EOSINT P380 SLS machine with exploded view of chamber.

Table 3.3 Levels of parameters

| S. No. | Parameter | Level of parameters | | | | |
		Level 1 (−2)	Level 2 (−1)	Level 3 (0)	Level 4 (1)	Level 5 (2)
1	Laser power (%)	50	60	70	80	90
2	Layer thickness (mm)	0.14	0.15	0.16	0.17	0.18
3	Scan speed (mm/s)	2800	2900	3000	3100	3200
4	Hatch spacing (mm)	0.3	0.4	0.5	0.6	0.7

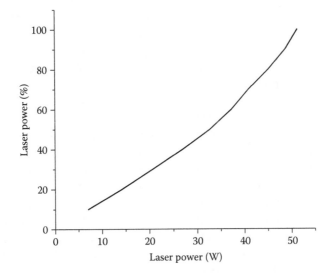

Figure 3.19 Variation of laser power with its percentage change.

Figure 3.20 Fabricated specimens using design of experiments.

3.6.2.2   Contribution of process factors on porous part strength

The fabricated specimens were tested for compression using a universal testing machine. As per ASTM 695D, cross travel speed of 1.3 mm/s was used for testing. Five specimens from each subgroup were tested for compressive strength, and average was considered, for further calculation. The combination of process parameters such as laser power (P%), layer thickness (L_t), scan speed (V), hatch spacing (H_s) and energy density (E) used for fabrication are enumerated in Table 3.4.

Table 3.4 Design of experiments with response

S. No.	P (%)	L_t (mm)	V (mm/s)	H_s (mm)	E (J/cm²)	$\sigma_{c(avg)}$ (MPa) ± σ
1	80	0.17	2900	0.4	3.88	7.125 ± 0.114
2	80	0.17	3100	0.4	3.64	7.256 ± 0.108
3	80	0.17	3100	0.6	2.42	6.599 ± 0.101
4	60	0.17	3100	0.4	3.00	8.013 ± 0.125
5	60	0.15	2900	0.6	2.14	5.632 ± 0.094
6	70	0.16	3200	0.5	2.56	7.455 ± 0.134
7	60	0.17	2900	0.4	3.22	7.282 ± 0.123
8	80	0.17	2900	0.6	2.59	6.989 ± 0.121
9	90	0.16	3000	0.5	3.25	7.025 ± 0.124
10	70	0.14	3000	0.5	2.73	6.539 ± 0.111
11	50	0.16	3000	0.5	2.18	6.244 ± 0.103
12	70	0.16	3000	0.5	2.73	8.202 ± 0.133
13	60	0.15	2900	0.4	3.22	6.680 ± 0.119
14	70	0.16	3000	0.5	2.73	8.024 ± 0.135
15	80	0.15	3100	0.4	3.64	7.627 ± 0.134
16	80	0.15	2900	0.6	2.59	7.392 ± 0.131
17	70	0.16	3000	0.5	2.73	7.737 ± 0.125
18	70	0.16	3000	0.7	1.95	6.595 ± 0.121
19	70	0.16	3000	0.3	4.54	7.925 ± 0.133
20	70	0.16	3000	0.5	2.73	7.965 ± 0.134
21	80	0.15	2900	0.4	3.89	7.316 ± 0.132
22	80	0.15	3100	0.6	2.42	7.669 ± 0.135
23	70	0.16	3000	0.5	2.73	8.001 ± 0.138
24	70	0.16	2800	0.5	2.92	6.703 ± 0.121
25	60	0.17	2900	0.6	2.14	5.526 ± 0.103
26	60	0.15	3100	0.4	3.00	7.450 ± 0.131
27	60	0.17	3100	0.6	2.00	6.012 ± 0.111
28	70	0.16	3000	0.5	2.73	7.865 ± 0.133
29	70	0.18	3000	0.5	2.73	6.636 ± 0.120
30	60	0.15	3100	0.6	2.00	6.780 ± 0.123
31	70	0.16	3000	0.5	2.73	7.987 ± 0.134

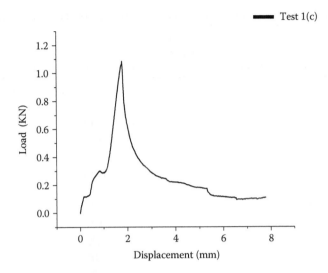

Figure 3.21 Load and displacement curves of test specimen.

The load and displacement curve of one typical test specimen 1(c) is shown in Figure 3.21. The subscript (c) used with test pieces shows positioning of the test piece in a subgroup of five from (a) to (e).

Using average compressive strength as a response factor, which is dependent upon selected process variables (predictors) were analysed for regression coefficient. On examination, the value of R-Sq (adj) was found to be the order of 94.1%. It represents good correlation among the data. The response was evaluated with respect to a significant coefficient having p value less than 0.05. Insignificant parameters having p value more than 0.05 were neglected. The ANOVA shown in Table 3.5 was checked for adequacy and lack of fit by means of F statistics at a confidence level of 95%. The model was established as adequate with insignificant lack of fit. The significant parameters and their interactions were used to prepare a regression model for predicting optimised compressive strength. The involvement of each significant parameter and its relation on part's strength was assessed.

The regression model stated in Equation 3.10 has been obtained to predict compressive strength of open porous polyamide mesh via significant parameters.

$$\sigma_c = -350 + 1.08P + 1255L_t + 0.141V + 21.4H_s - 184L_t \times H_s + 0.3P \times H_s$$

$$- 0.000175P \times V - 1.45P \times L_t - 0.000021V \times V - 3338L_t \times L_t \qquad (3.10)$$

$$- 16.6H_s \times H_s - 0.00322P \times P$$

Table 3.5 Analysis of variance for selected parameters and their interaction

S. No.	Source	DOF	Sum of square SS	Mean square MS	F value	P value	Remarks
1	P	–	–	–	–	0.000	R-Sq (Adj) (%) = 94.1
2	L_t	–	–	–	–	0.002	
3	V	–	–	–	–	0.006	
4	H_s	–	–	–	–	0.079	
5	$P \times P$	–	–	–	–	0.000	
6	$L_t \times L_t$	–	–	–	–	0.000	
7	$V \times V$	–	–	–	–	0.000	
8	$H_s \times H_s$	–	–	–	–	0.000	
9	$P \times L_t$	–	–	–	–	0.004	
10	$P \times V$	–	–	–	–	0.000	
11	$P \times H_s$	–	–	–	–	0.002	
12	$L_t \times V$	–	–	–	–	0.053	
13	$L_t \times H_s$	–	–	–	–	0.001	
14	$V \times H_s$	–	–	–	–	0.557	
15	Regression	14	15.1889	1.2657	34.70	0.00	$F_{0.05,14,16} = 2.37$,
16	Residual error	16	0.6566	0.0365	–	0.00	which is less than F value of regression. So model is adequate.
(a)	Lack of fit	10	0.3723	0.03723	1.81	0.241	$F_{0.05,10,16} = 2.49$, which is more than F value of lack of fit and p value is also more than 0.05. So lack of fit is insignificant.
(b)	Pure error	06	0.1232	0.02053	–	–	
17	Total	30	15.8456				

From the predicted model of strength it is observed that laser power contributes the greatest of 29%, hatch spacing 23%, layer thickness 16% and scan speed 11%. The contribution of each process factor involves their single and square terms. It is also observed that hatch spacing, which is correlated with energy density, contributes notably higher to the part strength in comparison to scan speed. The contribution of each significant parameter and their relations are shown in Figure 3.22.

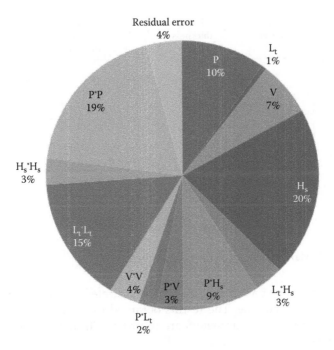

Figure 3.22 Contribution of significant process factors and interactions.

3.7 Discussion

3.7.1 Main effects

The main effect plots of the process variables are shown in Figure 3.23. It shows that an increase in laser power commencing 50%–70% causes an increase in porous part strength. Further increase in laser power commencing 70%–90% causes a decrease in part strength. This may be due to the reason of increase in energy density, which might have caused degradation of polyamide at this level. The trend of an increase in porous part strength remained and continued for layer thickness 0.14–0.16 mm while a decrease was observed for 0.16–0.18 mm.

These trends obtained in the plots strengthen the fact that net energy density is not adequate to sinter polyamide beyond 0.16 mm of layer thickness at a certain range of laser power. The increase in porous part strength has been observed when scan speed increases from 2800 to 3000 mm/s. The increase in scan speed further does not show a notable change in strength. With an increase of hatch spacing commencing 0.3–0.7 mm, the

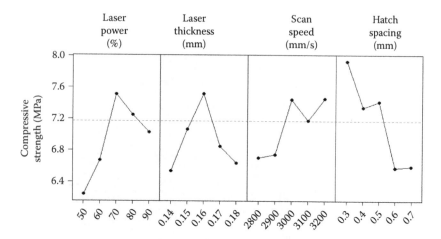

Figure 3.23 Main effect plot for linear parameters.

strength of porous part decreases, which indicates a decrease in overlapping of sintered volume. The results obtained for part strength in a certain range of process parameters are in order with the findings of other researchers for solid SLS parts [5,6,11].

3.7.2 Compressive strength maximization

Based on a regression model, within the range of a particular value of parameters, optimised values were obtained for maximum compressive strength using a trust region algorithm. Using optimised values of process factors given in Table 3.6, five pieces of open porous mesh were fabricated and tested for compressive strength. The average value was taken for assessment purposes. The values obtained in the experiment validated the developed model. The optimised values of theoretical compressive strength, experimental average compressive strength and difference are shown in Table 3.6.

Table 3.6 Value of parameters and comparison of results for compressive strength

% Laser power and (W) P	Layer thickness (mm) L_t	Scan speed (mm/s) V	Hatch spacing (mm) H_s	Energy density (J/cm²) E	Theoretical optimised compressive strength (MPa) σ_o(opt)	Experimental average compressive strength (MPa) σ_c(avg) $\pm \sigma$	% Difference with respect to mean
60% (37.3)	0.17	3100	0.3	4.01	8.98	8.85 $\pm$ 0.131	1.4

From the earlier, it is seen that the value of compressive strength found experimentally is in a small variance of 1.4% with the theoretical values. The results strengthen the validation of the developed model for optimized process variables.

3.8 Summary

The working of the SLS process has been studied in this chapter. The process parameters involved such as laser power, hatch spacing, scan speed, layer thickness and refresh rate have also been studied. The contribution of these process parameters in imparting the strength of SLS part in solid as well as in open cell porous regular interconnected structures have been studied. Optimised SLS process parameters have also been arrived at for polyamide (PA2200) OCPRIS, for enhancing the compressive strength without compromising its porosity. A regression equation was developed for predicting porous part strength and validated through experimental results. Laser power, hatch spacing and layer thickness have contributed more in porous part than scan speed. The interaction between hatch spacing and laser power has also been found on the higher side in contributing to porous part strength.

Revision questions

Q1. Explain the advantages of the Selective Laser Sintering Process.
Q2. Explain the working of the Selective Laser Sintering process.
Q3. Why does curling take place? How can it be prevented?
Q4. How is energy density related to laser power, scan speed and hatch spacing?
Q5. Explain the significance of the Andrew number.
Q6. What are the scanning paths to be followed in the SLS process? Explain with a neat sketch how hatching and contouring take place in the SLS process.
Q7. What is beam compensation? How it is applied in the SLS process?
Q8. What is skywriting? How it is used in the SLS process?
Q9. How does the positioning ability of the scanner help in removing unwanted errors in the SLS process?
Q10. What is the State Beer Lambert law of absorbance? How is the Gaussian curve for decay followed in the SLS process?
Q11. Explain in brief strength the related studies for solid SLS part.
Q12. What kind of foam structure can be fabricated through RP techniques? Explain fabrication of OCPRIS.
Q13. What significance can be drawn from the strength-related studies conducted by Singh et al. [21] for fabrication of OCPRIS of PA-2200?

Numerical problems

Q1. A selective laser system utilises laser power of 20 mW and spot size of 0.30 mm. If the speed of the laser is 150 mm/s, then calculate the average decay in laser power, laser exposure time and average density of laser exposure.

Ans1: The average decay of laser power is given by the relation.

$$H_{average} = \frac{P_L}{\pi W_0^2}$$

Here $P_L = 20$ mW

$$W_0 \left(\text{Half of the spot size}\right) = 0.30/2 \text{ mm}$$

So $H_{average} = \dfrac{20}{\pi \times 0.15^2}$ mW/mm^2

$$H_{average} = 28.29 \text{ W/cm}^2$$

The laser exposure time $t_e = \dfrac{2W_0}{V}$

$$t_e = \frac{2 \times 0.15}{150} \text{ mm-Sec/mm}$$

$$t_e = 0.002 \text{ Sec}$$

The average energy density ($E_{average}$) of laser exposure

$$E_{average} = H_{average} \times t_e$$

$$E_{average} = 28.29 \times 0.002 \text{ J/cm}^2$$

$$E_{average} = 0.05658 \text{ J/cm}^2$$

Q2. If the parameters for laser given in the aforementioned question remain the same and photon energy is 6.1×10^{-19} J/photon, calculate the number of photons per square centimetre area.

Ans2:

$$N_{ph} = \frac{E_{Average}}{E_{photon}}$$

$$N_{ph} = \frac{0.05658}{6.1 \times 10^{-19}} \text{ Photons/cm}^2$$

$$N_{ph} = 9.27 \times 10^{16} \text{ Photons/cm}^2$$

Q3. Polyamide 2200 (Nylon-12) was used to fabricate an open cell porous interconnected structure with 1 mm strut diameter and 60% inter-connected porosity using a Selective Laser Sintering machine. Laser power of 80% (60 W at 100%), layer thickness of 0.12 mm, scan speed of 2500 mm/s and hatch spacing of 0.25 mm were used as process variables. Calculate the compressive strength of the structure.

Ans3:

The given process variables are as under:

$$P = 80\%$$

$$V = 2500 \text{ mm/s}$$

$$H_s = 0.25 \text{ mm}$$

$$L_t = 0.12 \text{ mm}$$

The compressive strength of polymer (PA-2200) is given by

$$\sigma_c = -350 + 1.08P + 1255L_t + 0.141V + 21.4H_s - 184L_t \times H_s + 0.3P \times H_s$$
$$- 0.000175P \times V - 1.45P \times L_t - 0.000021V \times V - 3338L_t \times L_t$$
$$- 16.6H_s \times H_s - 0.00322P \times P$$

$$\sigma_c = -350 + 1.08 \times 80 + 1255 \times 0.12 + 0.141 \times 2500 + 21.4 \times 0.25 - 184 \times 0.12$$
$$\times 0.25 + 0.3 \times 80 \times 0.25 - 0.000175 \times 80 \times 2500 - 1.45 \times 80 \times 0.12 - 0.000021$$
$$\times 2500 \times 2500 - 3338 \times 0.12 \times 0.12 - 16.6 \times 0.25 \times 0.25 - 0.00322 \times 80 \times 80$$

$$\sigma_c = -350 + 86.4 + 150.6 + 352.5 + 5.35 - 5.52 + 6 - 35.6$$
$$- 13.92 - 131.25 - 48.0672 - 1.0375 - 20.608$$

$$\sigma_c = -5.1527 \text{ MPa}$$

The compressive strength of the structure is found to be negative. In order to ascertain the polymer-sintered behaviour, let us examine the sinter density of the structure using the following equation:

$$E = \frac{P}{V \times H_s}$$

Using a laser power graph between % and W, 80% laser power is equal to 45 W.

$$E = \frac{45}{2500 \times 0.25} = 7.2 \text{ J/cm}^2$$

The energy density achieved using the parameters stated in the problem is of the order of 7.2 J/cm^2 and clearly states that degradation of the polyamde-2200 took place, and the structure is unable to bear any type of compressive load.

Practice Calculations

Q1. A selective laser system utilises laser power of 30 mW and spot size of 0.20 mm. If the speed of laser is 250 mm/s, then calculate average decay in laser power, laser exposure time and average density of laser exposure.

Q2. If the parameters for laser given in the aforementioned question remain the same and photon energy is 6.1×10^{-19} J/photon, calculate the number of photons per square centimetre area.

Q3. Polyamide 2200 (Nylon-12) was used to fabricate an open cell porous interconnected structure with 1 mm strut diameter and 70% interconnected porosity using a Selective Laser Sintering machine. Laser power of 80% (60 W at 100%), layer thickness of 0.12 mm, scan speed of 2500 mm/s and hatch spacing of 0.25 mm were used as process variables. Calculate the compressive strength of the structure.

References

1. Noorani, R. 2006. *Rapid Prototyping: Principle and Applications*. John Wiley & Sons, Inc., Hoboken, NJ, pp. 1–377.
2. EOSINT P380. 2003. *Basic Training Manual*. EOS Gmbh, Munich, Germany.
3. Xie, J., Huang, S. and Duan, Z. 2005. Positional correction algorithm of a laser galvanometric scanning system used in rapid prototyping manufacturing. *International Journal of Advanced Manufacturing Technology*, 26(11):1348–1352.
4. Nelson, J.C., Xue, S., Barlow, J.W., Beaman, J.J., Marcus, H.L. and Bourell, D.L. 1993. Model of selective laser sintering of bisphenol: A polycarbonate. *Industrial & Engineering Chemistry Research*, 32(10):2305–2317.
5. Gibson, I. and Shi, D. 1997. Material properties and fabrication parameters in selective laser sintering process. *Rapid Prototyping Journal*, 3(4):129–136.
6. Thompson, D.C. and Crawford, R.H. 1997. Computational quality measures for evaluation of part orientation in free form fabrication. *Journal of Manufacturing Systems*, 16(4):273–289.
7. Williams, J.D. and Deckard, C.R. 1998. Advances in modelling the effects of selected parameters on the SLS process. *Rapid Prototyping Journal*, 4(2):90–100.
8. Miller, D., Deckard, C.R. and Williams, J.D. 1997. Variable beam size SLS workstation and enhanced SLS model. *Rapid Prototyping Journal*, 3(1):4–11.
9. Childs, T.H.C., Berzins, M., Ryder, G.R. and Tontowi, A. 1999. Selective laser sintering of an amorphous polymer–simulations and experiments. *Journal of Engineering Manufacture B*, 213:333–349.

10. Tontowi, A.E. and Childs, T.H.C. 2001. Density prediction of crystalline polymer sintered parts at various powder bed temperatures. *Rapid Prototyping Journal*, 7(3):180–184.
11. Chatterjee, A.N., Kumar, S., Saha, P., Mishra, P.K. and Choudhary, A.R. 2003. An experimental design approach to selective laser sintering of low carbon steel. *Journal of Materials Processing Technology*, 136:151–157.
12. Dewidar, M.M., Dalgarno, K.W. and Wright, C.S. 2003. Processing conditions and mechanical properties of high-speed steel parts fabricated using direct selective laser sintering. *Journal of Engineering manufacture*, 217:1651–1663.
13. Storch, S., Nellessen, D., Schaefer, G. and Reiter, R. 2003. Selective laser sintering: Qualifying analysis of metal based powder systems for automotive applications. *Rapid Prototyping Journal*, 9(4):240–251.
14. Ning, Y., Fuh, J.Y.H., Wong, Y.S. and Loh, H.T. 2004. An intelligent parameter selection system for the direct metal laser sintering process. *International Journal of Production Research*, 42(1):183–199.
15. Ning, Y., Wong, Y.S. and Fuh, J.Y.H. 2005. Effect and control of hatch length on material properties in the direct metal laser sintering process. *Journal of Engineering Manufacture*, 219:15–25.
16. Zarringhalam, H., Hopkinson, N., Kamperman, N.F. and De, J.J. 2006. Effects of processing on microstructure and properties of SLS nylon 12. *Materials Science and Engineering*, 4:172–180.
17. Ajoku, U., Hopkinson, N. and Caine, M. 2006. Experimental measurement and finite element modelling of the compressive properties of laser sintered nylon-12. *Materials Science and Engineering*, A-428:211–216.
18. Caulfield, B., McHugh, P.E. and Lohfeld, S. 2007. Dependence of mechanical properties of polyamide components on build parameters in the SLS process. *Journal of Materials Processing Technology*, 182:477–488.
19. Ashby, M.F., Evans, A.G., Fleck, N.A., Gibson, L.J., Hutchinson, J.W. and Wadley, H.N.G. 2000. *Metal Foam: A Design Guide*. Butterworth Heinermamm Press, Boston, MA, pp. 1–251.
20. ASTM D695D. 2015. Standard test method for compressive properties of rigid plastics.
21. Singh, J.P., Pandey, P.M. and Verma, A.K. 2016. Fabrication of three dimensional open porous regular structure of PA-2200 for enhanced strength of scaffold using selective laser sintering. *Rapid Prototyping Journal*, 22(4):752–765.

chapter four

Fused deposition modelling

Applications and advancements

Kamaljit Singh Boparai, Rupinder Singh
and Jasgurpreet Singh Chohan

Contents

4.1   Introduction to FDM

The emergence of rapid prototyping techniques revolutionized the modern industry by producing parts having complex shapes and meeting fast-changing customers' requirements within a stipulated time period. The current rapid prototyping technologies are limited to fabricated prototypes with a small range of materials. The production rate of these prototypes is also less. In order to reduce the production time and cost, a multistep procedure has to be adopted. This multistep procedure is termed as rapid tooling. The rapid tooling offers to produce parts with a wider variety of material and in large quantities.

The traditional prototyping methods are usually time consuming and very expensive. Moreover, the accuracy of prototypes depends upon the skill of the technician. Due to this, very small iterations in the product design are possible, which further increases the final product cost. Additionally, low-volume products and rapidly changing high-volume products required quick and cost-effective development procedures to be able to compete in the market. Many researchers and manufacturers tried to explore different rapid prototyping techniques, and this led to a major shift from traditional prototyping tooling practices to rapid prototyping tooling options. Many researchers realized that tools produced by rapid tooling are more durable and accurate, but there is also a scope for them to be employed in the production process. There are a number of rapid prototyping techniques that are available, but the additive manufacturing (AM) technology has emerged as a solution to shorten the product development cycle, achieve flexibility for manufacturing small batch sizes and carry out manufacturing of complex designed components at a low cost.

Fused deposition modelling (FDM) works on an additive manufacturing principle and is most widely used for modelling, prototyping and the production of customer-specific applications. It is one of the techniques used for 3D printing of parts. According to Stratasys, 'FDM Technology works with production-grade thermoplastics to build strong, durable and dimensionally stable parts with the best accuracy and repeatability of any 3D printing technology'. FDM works on an 'additive' principle by laying down material in layers; a plastic filament or metal wire is unwound from a coil and supplies material to produce a part.

4.2   History of fused deposition modelling

Additive manufacturing, or 3D printing, has been commonly used for building prototypes since the 1980s and today has become the fastest growing, most affordable, ecofriendly way to fabricate consumer goods. The technology was developed and invented by S. Scott Crump in the late 1980s, and after that it was commercialized in 1990. Mr Scott is the cofounder and chairman of Stratasys, Ltd., a leading manufacturer of 3D printers. The term 'fused deposition modeling' and its abbreviation to FDM are trademarked by Stratasys Inc. Besides this, a number of organizations have since adopted similar technologies of 3D printing under different names. The Brooklyn-based company MakerBot (now owned by Stratasys) was founded on a nearly identical technology known as Fused Filament Fabrication (FFF). FFF is equivalent to FDM and was coined by members of the Rep Rap project to give a phrase that would be legally unconstrained in its use. It is also sometimes called Plastic Jet Printing (PJP).

Many companies that manufacture FDM printers also offer a range of 3D printing services to clients, including external 3D modelling and printing. The popularity of FDM in a variety of industries ranges from automotive (BMW, Hyundai, Lamborghini) to consumer goods manufacturing (Black & Decker, Dial, Nestle). These companies use FDM throughout their product development, prototyping and manufacturing processes. FDM also caters to the need of low-volume tooling and durable end-use parts for industries such as aerospace, defence, architecture, consumer products, entertainment and medical. Furthermore, FDM continuously appeals to large manufacturers, designers, engineers, educators and other professionals.

4.3   Working of fused deposition modelling

The FDM process is very simple as it consists of three phases: pre-processing, production and post-processing. In pre-processing, the FDM process starts with the input of computer-aided design (CAD) files. Before the execution of printing, its CAD file must be converted to a format that a 3D printer can understand. Usually, the CAD file is converted into STL format. As already mentioned, FDM is a computed automated manufacturing process, which generally takes the data from a CAD file. The software package 'slices' the CAD model into a number of thin layers. The production of parts starts with movement of the machine head. Based upon the path generated by computer software, the controller unit directs the machine head to build the part model with layer-by-layer deposition of material (ABS) up one atop another. A feedstock filament of recommended standard material is fed by rollers through the liquefier head and extruded through a nozzle. The system is designed to operate in a temperature-controlled environment to maintain sufficient fusion energy between each layer. The various processing steps of the FDM process are shown in Figure 4.1.

FDM printers use two kinds of materials, a modelling material, which constitutes the finished object, and a support material, which provides necessary support to the object as it is being printed.

During printing, these materials take the form of plastic threads or wires but generally known as filaments, which are unwound from a coil and fed through an extrusion nozzle. The nozzle placed in the liquefier head melts the filaments and extrudes them onto a base, sometimes called a build platform or table. Both the nozzle and the base are controlled by the computer software that translates the dimensions of an object into X, Y and Z coordinates for the nozzle and base to follow during printing.

In a typical FDM system, the extrusion nozzle moves over the build platform horizontally (XY plane), 'drawing' a cross section of an object onto the platform (Figure 4.2). This thin layer of plastic cools and hardens, immediately binding to the layer beneath it. Once a layer is completed, the base is lowered (Z direction). Usually, it moves by about one-sixteenth of an inch – or it depends upon the layer thickness – and makes a space

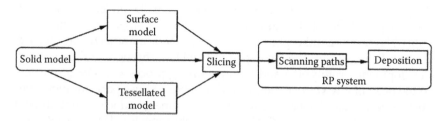

Figure 4.1 FDM process.

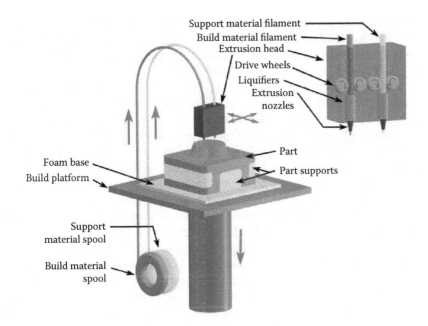

Figure 4.2 Schematic of FDM process.

for the next layer during the part building process. Based upon the part geometry, the process continues until the part builds completely.

Printing time depends on the geometry (shape and size) of the object being manufactured. Small objects having a few cubic millimetres and tall, thin objects print quickly, while larger, more geometrically complex objects take more time for printing. A part of this printing time also depends upon part density (solid, single sparse or double sparse). The production rate of FDM is fairly low as compared to other 3D printing methods, such as Stereolithography (SLA) or selective laser sintering (SLS). Finally, various post-processing techniques can be used to break away support material and improve the surface quality of parts.

4.4   Fused deposition modelling materials

FDM technology uses only recommended thermoplastics that are tried and tested by the experts, and the same have applications in various traditional manufacturing processes. The FDM thermoplastics also meet the requirements such as tight tolerances, toughness and environmental stability or specialized properties such as electrostatic dissipation, translucence, biocompatibility, VO flammability or FST ratings. The most common printing material for FDM is acrylonitrile butadiene styrene (ABS), a common thermoplastic that is used to make many consumer products, from LEGO bricks to white water canoes. Along with ABS, some FDM machines also print in other thermoplastics, such as polycarbonate (PC) or polyetherimide (PEI). Support materials are usually water-soluble wax or brittle thermoplastics, such as polyphenylsulfone (PPSF).

Thermoplastics can endure heat, chemicals and mechanical stress, which make them an ideal material for printing prototypes that must withstand testing. And because FDM can print highly detailed objects, it is also commonly used by engineers who need to test parts for fit and form.

FDM is also used to produce end-use parts, particularly small, detailed parts and specialized manufacturing tools. Some thermoplastics can even be used in food and drug packaging, making FDM a popular 3D printing method within the medical industry. These days, wide varieties of materials are available for the FDM process, which extends its application range in aerospace companies, medical device makers and limited-production automakers. The FDM materials can be classified as standard materials and application-specific materials (Figure 4.3). A part of a powerful range of additive manufacturing materials, including clear, rubberlike and biocompatible photopolymers and tough high-performance thermoplastics are also used to maximize the benefits of 3D printing throughout the product development cycle.

4.4.1   Standard materials

The standard materials of filament for the FDM process are acrylonitrile butadiene styrene (ABS), polylactic acid (PLA), polycarbonate, polyamides, polystyrene, polyethylene and polypropylene. These materials are used for their high strength and heat-resistant properties. ABS is a common end-use engineering material to produce functional prototypes, and its various available grades are summarized in Table 4.1.

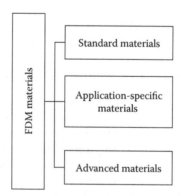

Figure 4.3 FDM materials.

Table 4.1 Different grades of ABS

S. No	Base material	Grade
1	Acrylonitrile butadiene styrene	ABS plus
2	Acrylonitrile butadiene styrene	ABS-M30
3	Acrylonitrile butadiene styrene	ABS-M30i
4	Acrylonitrile butadiene styrene	ABSi
5	Acrylonitrile butadiene styrene	ABS-ESD7
6	Nylon 6	FDM Nylon 6
7	Nylon 12	FDM Nylon 12
8	Nylon 12	FDM Nylon 12CF
9	Polycarbonate acrylonitrile butadiene styrene	PC-ABS
10	Polycarbonate	PC
11	Polycarbonate	PC-ISO
12	Polyphenylsulfone	PPSF/PPSU
13	FDM thermoplastic material	ULTEM 1010
14	FDM thermoplastic material	ULTEM 9085

4.4.2 Application-specific materials

The use of RP technology for RT and RM has given rise to the development of application-oriented composites. FDM is capable of yielding strong composite parts as bond forms between successive roads and layers due to partial and full melting of composite feedstock filaments [1]. Work has been in progress in some universities and research institutions to develop new metallic and ceramic materials for rapid fabrication of functional components by FDM with higher mechanical, thermal and wear-resistant properties [2].

The various materials such as wax, paper, nylon, glass-filled nylon, metal-filled nylon, metal-filled ABS and ceramics have been used. Rutgers University in the United States has carried out considerable work in the development of fused deposition of ceramics (FDC) and metals. It has used the process to fabricate functional components of a variety of ceramic and metallic materials such as silicon nitrate, PZT, aluminium oxide, hydroxyapatite and stainless steel for a variety of structural, electro-ceramic and bio-ceramic applications. Researchers at Virginia Tech have developed a new high-performance thermoplastic composite for FDM that involves thermotropic liquid crystalline polymers (TLCP) fibres (by heating a polymer above its glass or melting transition point), and they have used it in an FDM system to fabricate prototype parts.

The tensile modulus and strength of this material were approximately four times those of ABS. The feedstock filament for FDM was prepared with the mixture of polypropylene (PP) and ceramic powder such as mullite ($3AL_2O_3$, $2SiO_2$), fused silica (SiO_2), titanium dioxide (TiO_2) and alumina (Al_2O_3). The various additives such as tackifier, elastomers, plasticizer and wax are added to control the various properties such as flexibility, stiffness, viscosity and strength of filament [3].

The use of polylactic acid (PLA) and tricalcium phosphate (TCP) as a biodegradable composite is state of the art in tissue engineering and maxillofacial surgery. The general suitability of PLA for the processing with FDM was evaluated, and material-specific effects (e.g. crystallization and shrinkage) were observed, and the characterization of the semi-crystalline biodegradable material by thermal, mechanical and microscopic analysis was carried out. Components from PLA/TCP, which have sufficient mechanical properties for their potential use as scaffoldings, were obtained. The feedstock filament for FDM was prepared with a composite of Polyamide 6 (PA 6) or Nylon 6 filaments reinforced with 0.5 and 1.0 wt.% α-Al_2O_3 nanoparticles. Tensile tests on single filaments have demonstrated that the average yield strength and Young's modulus of 0.5 and 1.0 wt.% α-Al_2O_3 nanoparticles reinforced PA 6 filaments are greatly improved in the range of 3%–173% and 9%–90% as compared to those of neat PA 6 filament, respectively. The thermal stability and mechanical strength of as-fabricated composite filament is more than neat PA 6 filament [4].

A new polymer matrix composite (PMC) feedstock material consists of iron powder filled in an ABS, and surfactant powder (binder) material was investigated experimentally. The effect of higher powder loading of iron filler affected the hardness, tensile and flexural strength of PMC material. The tensile properties of Polycarbonate (PC) material used in the FDM systems have been affected by the FDM process parameters such as air gap, raster width and raster angle. The results show that FDM-made parts have tensile strength in the range of 70%–75% of the moulded and extruded PC parts. The FDM technology offers the potential to produce the functional parts with a variety of materials including composite materials and have application to direct rapid tooling. The composite that consists of iron particles in nylon are investigated experimentally, and feedstock filament of this composite has been produced on a screw extruder and used directly without changing any software and hardware of the FDM system. The composite parts have higher mechanical and thermal properties than ABS parts [5].

A new polymer nanocomposite material – which is a mixture of polycaprolactum (PCL), montmorillonite (MMT) and hydroxyapatite (HA) as a filler – can be used as an alternative material for FDM. PCL is bioresorbable, more stable in ambient condition, less expensive and easily available as compared to PLA. For FDM applications, the biomaterials such as PCL, PP-β-TCP, POT/PBT, PCL-β-TCP and β-TCP have been used till now. Various medical-grade materials are also available, which can be used to fabricate RP models on the basis of their use in different medical applications [6].

4.4.3 Advanced materials

Apart from the aforementioned materials for the FDM system, many researchers have recently focused on the development of 4D materials. The 4D printing utilizes multi-materials that have a capability of shape transformation from one state to another and can takes place directly off the print bed. The shape transformation over time is possible upon exposure to certain stimuli such as pressure, wind, water, heat, light, or simple energy as an input. In particular, 4D printing is the extension of 3D printing with one or more additional design dimensions, such as material gradation over distance or direction, response or adaptation over time, or controlled anisotropy throughout volume. In other words, the 4D printing process has an added dimension – that is, time.

The 3D printing process and the combination of smart materials has led to the development of a new technology called 4D printing. The smart materials consist of shape memory polymers (SMPs) that react to an applied stimulus in order to recover large strains (in longitudinal and transverse directions) and allow for the fabrication of small parts in a wide variety of geometries. However, SMPs are easier to process, lightweight and inexpensive as compared to their metallic counterparts, shape memory alloys. Yang et al. [7] explored the property of SMP for the design and fabrication of a modular omnidirectional joint in a robotic arm with variable stiffness. Figure 4.4a illustrated the primitive bending of the part with time. This part design is composed of bars and disks. The disks in the centre act as stoppers. By adjusting the distances between the stoppers, it is possible to set the final bending angle. Figure 4.4b shows the fabrication of a time-varying curve.

4D printing encompasses various potential applications such as:

1. *Robotic features*: The parts fabricated by 4D materials can behave like robots without depending upon the electromechanical devices. The 4D prints can be used as actuators and controllers for various industrial automations.
2. *Garments*: The printing of clothes with 4D materials provides maximum comfort due to its response to weather conditions.
3. *Orthopaedic applications*: The 4D parts are are more compatible with a body (human or animal) than rigid metal supports.
4. *Military applications*: The 4D printing process promotes a self-assembly feature and parts that can be used for the automatic assembly of weapons and tools for the army.

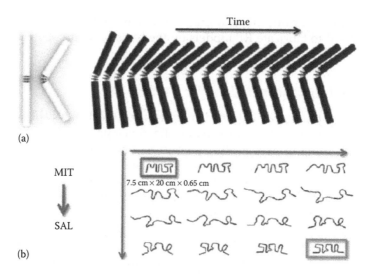

Figure 4.4 4D material bending process (a) Bending process with time change (b) MIT impression with 4D material. (From Choi, J. et al., *3D Print. Addit. Manuf.*, 2, 159–167, 2015 [8].)

4.5   *Rapid tooling with FDM process*

The FDM process has been commercially used as a rapid tool in rapid casting (RC) applications such as sand, vacuum and investment casting and injection moulding processes. Figure 4.5 shows the rapid tooling with the FDM process for RC applications in the form of a flowchart. The selection of the appropriate rapid tooling route from various direct (FDM, LOM, SLA, SLS) and indirect routes (i.e. vacuum casting and investment casting) is an important decision-making process in sand and investment casting. A methodology has been demonstrated and validated with industrial applications and is a robust decision-making tool for selection of the appropriate tooling route. The term 'rapid investment casting' represents the employment of RP and RT techniques in IC. The cost involved in designing and fabrication of metal tooling for a wax injection process can be overcome by using AM techniques to fabricate sacrificial patterns for IC and reduces the overall lead time involved in production of prototype casting with excellent quality [9].

Moreover, IC patterns fabricated by RP processes (FDM, MJM) are resistant to heat, humidity and post curing. Omar et al. [10] evaluated the quality characteristics such as surface roughness, dimensional accuracy and pattern shrinkage of IC patterns fabricated by FDM, 3D printing and MJM and found that FDM and MJM were more suitable as an RT for IC applications. By employing AM-fabricated patterns to produce the prototypes, there is no need to commit to production tooling for a single part or

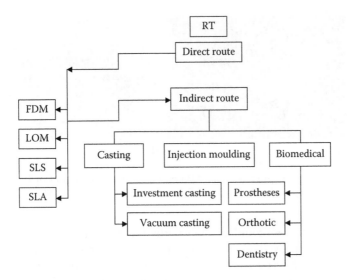

Figure 4.5 Rapid tooling by FDM process.

small quantity production. RP models can be used to create male models for tooling, such as silicone rubber moulds and investment casts, and their applications can be grouped into design engineering, analysis and planning, and tooling and manufacturing. Free-form fabrication increases manufacturing flexibility by eliminating the need for part-specific tools and can be used for direct manufacturing or to make tooling for parts [11].

In injection moulding applications, rapid tooling methods have certain limitations in terms of mould material, accuracy, surface finish and mould life. The CAD-based RT selection process for a given injection mould requirement was evaluated to identify critical features that could be modified to improve manufacturability, better quality, lower cost and shorter lead time. The studies also focused on the development of new polymeric-based material in order to cater applications in the injection moulding process. The development and testing of a new metal-polymer composite material for use in the FDM process increases the application range of rapid tooling in the injection moulding process.

FDM has been commonly used to fabricate tissue-engineering scaffolds and provides a fast and cost-effective process to develop more suitable biomedical products. Customized patterns of prostheses can be made by using the AM process, and finally cast components can be prepared using an investment casting process. A new continuous tool-path strategy for an open-source FDM machine and a new integrated tool to automatically produce customized products were designed. A customized silicone tracheal stent was fabricated on an FDM machine to have good surface quality, overall fabrication efficiency and reduced costs. In prosthetic, orthotic and dental applications,

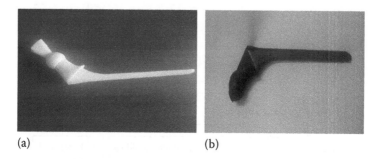

(a)   (b)

Figure 4.6 Hip joint prepared by FDM-based IC: (a) ABS model of hip joint and (b) Final cast component. (From Singh, R. et al., *Procedia Mater. Sci.*, 6, 859–865, 2014 [12].)

an RM or RT process could facilitate better record keeping, streamlined processing and improved access to expert care for people in remote locations. Figure 4.6a and b respectively, show an ABS model of a hip joint prepared via FDM and the final cast component prepared with IC (as indirect route).

4.6   Surface finish of FDM parts

In FDM, the slicing and conversion of the CAD model deteriorates the surface finish and dimensional accuracy of the parts. The Standard Triangulation Language (STL) format approximates the part surface as a web of triangles by tessellating the CAD model. It simplifies the geometry, but the part loses its resolution as after tessellation only triangles (not curves) represent its outline. This causes chordal error or defect (Figure 4.7), which is a difference between the original CAD surface and the corresponding triangle of the tessellated model [13]. The major defect that arises from the chordal effect is the dimensional variation in FDM parts. The actual dimensions as compared to the original CAD model vary non-uniformly on different locations depending

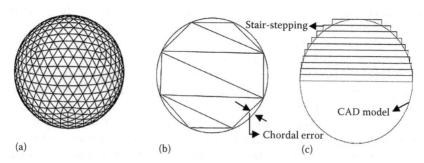

(a)   (b)   (c)

Figure 4.7 Model of sphere: (a) Tessellated CAD Model; (b) Chordal error after tessellation and (c) Stair stepping after slicing. (From Chohan, J.S. and Singh, R., *Rapid Prototyping J.*, 23, 495–513, 2017 [15].)

upon part geometries. Generally, the small-sized parts and intricate shapes are found to be oversized, having lesser dimensional accuracy while shrinkage occurred in large-sized parts. The simple solution to this problem is to do positive offset to the outer surface. However, this would increase the part dimensions, which would require precise post-processing and it is still very difficult to attain original dimensions [14]. The staircase (stair-stepping) effect is due to the slicing method as each layer (slice) of molten plastic is laid on the previous slice to complete the part that creates predictable roughness on the outer surface (Figure 4.7). Other defects may be accounted as start/stop errors (seams), support structure burrs and low-dimensional accuracy. These are the inherent defects and cannot be completely eradicated, which ensured poor surface characteristics of the FDM parts.

The factors that influence the surface quality of FDM parts are as follows:

- Stair-stepping effect
- Accuracy of machine
- Material flow rate
- Shrinkage and residual stresses due to resign solidification and cooling
- Support material marks on parts

The surface roughness parameters of fused deposition modelling prototypes can be predicted by a profilometric, microscopic and SEM analysis. The techniques developed to improve surface finish are generally classified under two categories – that is, pre-processing and post-processing (Figure 4.8). All surface refinement methods adopted before fabrication of

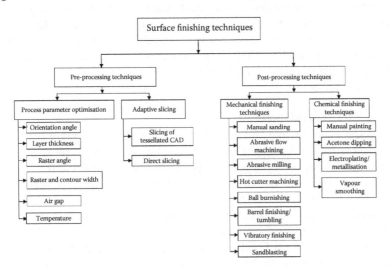

Figure 4.8 Classification of surface finishing techniques. (From Chohan, J.S. and Singh, R., *Rapid Prototyping J.*, 23, 495–513, 2017 [15].)

the FDM parts are classified under pre-processing while in post-processing, the parts are treated after extrusion under the nozzle is completed.

4.7   Pre-processing techniques

Since the inception of the FDM process, extensive research has been done to improve the surface finish of FDM parts, but it is inevitable that some degree of surface roughness exists. Many researchers have been successful to control and reduce the surface roughness using different pre-processing and post-processing techniques. Broadly, the parameters affecting the performance of the FDM process are divided in geometry-specific parameters, operation-specific parameters, machine-specific parameters and material specific parameters [14]. The latter two are not very effectual because material and machines are already optimized for best output and their parameters are difficult to alter afterwards at the production stage. The addition of aluminium powder and epoxy resins as filler in ABS wire may induce significant improvement in surface finish. But this could resist the use of FDM prototypes for specific applications where ABS composition plays vital role. Thus, researchers have focused on optimizing the geometry- and operation-specific parameters, which are discussed in the next section.

4.7.1   Process parameter optimization

In spite of the fact that surface roughness cannot be completely eliminated, many researchers have tried to reduce the defect by executing the process at different parameter levels and modifying the slicing techniques. The variation in various process parameters significantly affects the surface finish and mechanical properties of parts; thus, many researchers have performed optimization studies. In pre-processing techniques, the different process parameters of the FDM machine are optimized and discussed in the next sections.

4.7.1.1   Build orientation

This is the most flexible and notable pre-processing parameter studied by researchers for attaining the best surface characteristics. The angle of the orientation or deposition angle can be altered with respect to the machine coordinate system in the CAD model (Figure 4.9) to achieve the desired objectives. Generally it is focussed on minimum cost, time, support, material usage and most importantly the surface finish required on the specific surface. Kattethota and Henderson [16] initially investigated the effects of build orientation on FDM parts and found the angle of 0° yielded the maximum surface finish. In most of the studies, the orientation angles of 0° and 90° were found to be most effective for surface finish, build times and cost. On the other hand, build angles between 40° to 60° decreased

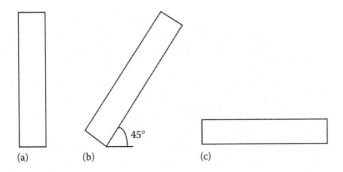

Figure 4.9 Side view of rectangular test parts different orientation angles (a) 90°; (b) 45° and (c) 0°. (From Chohan, J.S. and Singh, R., *Rapid Prototyping J.*, 23, 495–513, 2017 [15].)

the surface finish and increased production cost as the maximum support material was used due to tilt of model.

The researchers have tried to optimize build times and surface finish by orienting the models at different X and Y directions using genetic algorithm and multi-criteria decision-making. A few authors observed an orientation angle of 30° as the most effective for surface finish and mechanical properties such as flexural and impact strength through Grey Taguchi optimization techniques. In another study, the test parts were fabricated via FDM by varying angles from 0° to 90° on a single part and got minimum surface roughness at 90°. Studies suggested that the horizontal surface has a better surface finish than the vertical surface and the worst for circular shapes due to the formation of elliptical curves on the downfacing surface. Some authors found that part dimensions showed minimum variation at 0° and 30° orientation angles in the case of different shapes of test parts.

In most of the cases discussed, the optimization study was conducted by fabricating standard square and rectangular models oriented at different angles. The practical problems would arise when actual parts were needed to be manufactured having intricate shapes and minute details. Moreover, in the earlier cases, the goal is to achieve best surface finish on a pre-specified surface and but for rapid tooling and casting applications, the surface finish is imperative on all the surfaces and locations. Thus, it is deceptive to conclude that surface finish can be optimized at a specific orientation angle on all the locations of the FDM part.

4.7.1.2   Layer thickness

Layer thickness is the height of the layer (slice) deposited by the extrusion nozzle and is a highly influential process parameter affecting the part surface quality. The thickness of the bead extended from the nozzle tip or displacement in the z-axis between successive layers deposited on the bed determines the height of the stair-step. Generally, decreasing the nozzle diameter decreases the stair-stepping, but this increases build time and cost [13]. Thus, there must be a trade-off between build time and layer thickness before initiating the process, as reducing build time would increase stair-stepping, which would further require post-processing. It can be concluded that the best surface finish and dimensional accuracy were achieved from minimum layer thickness.

There is a minimum limit of layer thickness beyond which it cannot be decreased, which limits the domain of this parameter. Moreover, this strategy is efficient where layer thickness of the FDM machine can be varied. However, few machines do not have the option of variable layer thickness, which led to optimization studies of a few other effective parameters.

4.7.1.3 Raster width and contour width

Raster or road width is the width of the path of molten bead that is deposited on the base plate (Figure 4.10), and it is 1.2–1.5 times the nozzle diameter depending upon material [14]. Generally, road width is closely related to nozzle tip diameter and varies linearly [13] with it. The impact of variation is the same as layer thickness as the minimum raster width resulted in maximum surface finish and dimensional accuracy. It has been noticed that smaller road widths also enhanced mechanical properties of FDM parts. Contour width is the width of the outermost of the uppermost or outermost layer of the part built by FDM. The direction and fill strategy of contour and raster width depends upon tool paths generated by the FDM machine and strongly affects surface and mechanical characteristics. The wider contour width is best for surface finish and dimensional accuracy as thin contours are easily deformed by heat evolved during extrusion [17].

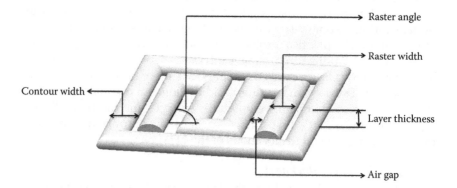

Figure 4.10 Various FDM pre-processing parameters. (From Chohan, J.S. and Singh, R., *Rapid Prototyping J.*, 23, 495–513, 2017 [15].)

4.7.1.4 Air gap

The gap between two adjacent roads on the same layer is road gap or air gap as depicted in Figure 4.10. Generally, the air gap has taken zero as the default, which results in the touching of ends of the two nearest beads. The air gap can be increased (positive gap) in order to reduce build time, and density of parts reduces. On the other hand, overlapping of two roads resulted in a negative gap, which led to the dense structure having a longer build time. The surface finish generally increases with both negative and positive air gaps while minimum surface finish and dimensional accuracy occur at zero air gaps [15].

4.7.1.5 Raster angle

Raster angle is the direction of raster relative to the x-axis of the build table (Figure 4.10). There are generally two raster angle strategies adopted by researchers – that is, cross (0°/90°) and criss-cross (−45°/+45°). Some authors have varied the raster angle by 30°/60° for optimization of different outputs. The setting of rasters plays a critical role in bonding and strength of the inner layers and surface finish of the upper layer. Kumar et al. [18] concluded that −45°/+45° raster angle styling renders a maximum surface finish. Moreover, a raster angle of 30° provides maximum flexural and impact strength. The variation in interpretations of different authors may be due to difference in CAD models and other parameters.

4.7.1.6 Model temperature

The temperature at which heaters are maintained to melt the outgoing ABS wire before extrusion is termed as model temperature. Its variation could change the fluidity of material, which further affects the surface roughness. The higher temperature produces a smooth surface due to a delay in solidification. The viscosity of semi-molten ABS plastic material is less, which promotes rounding off the stair-steps leading to a better surface quality and bonding strength. Vasudevarao et al. [13] recognized model temperature to be the third most significant factor affecting surface finish after layer thickness and orientation angle. Generally, the use of slightly higher extrusion temperatures would result in better surface quality. At the chamber temperature below 80°, there was less adhesion of plastic with the base plate while dimensional accuracy decreased above 100°. The lower temperature reduces the surface finish but prevents adhesion of part material with the base plate and thus, parts are detached easily after fabrication. On the other hand, higher temperature leads to good surface finish but poor dimensional accuracy due to a higher flowability of plastic material.

The head temperature is set according to the type of support material and model material to be processed. The latest version of FDM can accommodate 12 different types of model materials, and each material has its own processing temperature. The FDM liquefier head can be set at a particular temperature for the processing of a particular material. The machine automatically adjusts the liquefier head temperature according to the selected model material.

Generally, there are three possible interior fill types – medium, high and solid – in order of increasing material usage and build times. If enhancement of tensile strength and surface hardness is not a primary objective, then nearly the same surface characteristics are obtained in all interior fill strategies. These strategies are built by machine based upon CAD and STL files, which may vary with different geometries and FDM machines.

The aforesaid parameters significantly exert maximum influence on surface finish though many other parameters that have been studied, but their impact was trivial. The variation of support and part material, feed rate, extrusion speed, material colour, part geometry, etc., has been studied; however, these evolved to be the least important for surface finish. The advanced optimisation tools showed raster width as the most significant for top surface while orientation angle and layer thickness for the bottom and side surfaces, respectively.

The major drawback of the optimization of process parameters for better surface finish and dimensional accuracy is the fear of losing few mechanical properties of FDM parts. The optimized setting of different parameters varies with part geometry and manufacturing environment, and thus different algorithms are proposed by authors to attain the best surface characteristics. Thus, it is difficult to optimize the parameters every time for different components due to their varying geometry during the production phase.

4.7.2   Adaptive slicing

The most significant barrier against fabrication of high surface quality parts seemed to be longer production times due to a decrease in layer thickness, which further increases manufacturing cost. Thus, adaptive slicing techniques have been developed, which act as a balancing agent between surface finish and build times. It has been observed that constant layer thickness leads to a wastage of time in some situations as the layer (slice) thickness does not affect surface roughness. This is completely case specific depending upon dimensions and shape of parts. The adaptive slicing focuses on developing such algorithms, which can automatically vary slice height and reduce production time and generate variable tool paths depending upon part geometry to achieve minimum surface roughness and production time. Originally, the solid model is sliced by CAD software before transferring the STL data to the FDM machine. The slicing of the CAD model is carried out either directly on a solid model or after tessellation. In slicing, the set of horizontal planes divide the CAD model in closed curves or polygons, and the space between two successive horizontal planes is called a slice.

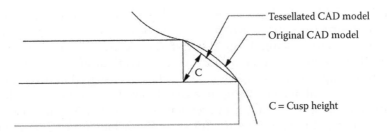

Tessellated CAD model

Original CAD model

C

C = Cusp height

Figure 4.11 Cusp height due to stair stepping in FDM test part. (From Pandey, P.M. et al., *Rapid Prototyping J.*, 9, 274–288, 2003 [20].)

4.7.2.1 Slicing of tessellated CAD model

The cusp height adjudicates the layer thickness, which further controls the surface roughness of FDM parts (Figure 4.11). The maximum limit of cusp height is specified by the user, and thus algorithm automatically slices the CAD file within a given range. This method imposes restrictions to surface roughness and cusp height uniformly for the whole part surface [19].

Mani et al. [19] proposed region-based adaptive slicing where the user is free to vary surface roughness at different locations of the part surface as per requirement. Researchers used a stepwise uniform refinement technique where initially the CAD model undergoes uniform slicing, and then each slice is again subdivided into smaller slabs of different heights by interpolation. The problem of implementation of this technique in industry was solved in a manner where different parts were sliced independently based upon part features and design. There are few limitations in the aforementioned techniques as complexities and errors in tessellated CAD files further need software to cure.

4.7.2.2 Direct slicing

These aforementioned limitations are eliminated in direct slicing where software directly slices the CAD file without STL conversion, which is accurate, fast and requires less storage memory. The file is stored as analytical surfaces such as rectangular, trapezoidal and parabolic. However, there is difficulty in orientation of the model afterwards as the CAD model is stored as mathematical definitions and analytical surfaces rather than points and co-ordinates. The CAD file is directly fed to an FDM machine, avoiding the tessellation phase. Bordoni and Boschetto [21] proposed a method to mathematically generate thick solid surfaces to achieve better surface characteristics in shorter times without using repair interventions as a faultless model directly goes for manufacturing. The software and hardware were developed for implementing curved layer-slicing algorithms with an aim to enhance both mechanical properties and surface finish of FDM parts. The curved layer-slicing method resulted in better material structure and mechanical strength due to fibre continuity as against flat layer slicing. It could be interpreted from the earlier discussions that adaptive slicing is a cumbersome technique of developing algorithms to improve the surface finish of industrial parts. Moreover, surface finish can be enhanced only up to a certain level, depending upon hardware and software, which is unacceptable to make FDM parts used for precision rapid tooling. The modification in various processing parameters and slicing algorithms could not outstretch the finishing level as compared to conventional or CNC machining.

4.8   Post-processing techniques

The present context realizes various post-processing techniques for surface finishing employed after the production phase. These are generally categorised into two types – mechanical finishing and chemical finishing techniques, which are addressed in the next section.

4.8.1   Mechanical finishing techniques

As the name depicts, these techniques concentrate on mechanical cutting or pressing the peaks of surface profiles. Generally, the techniques are replicated from conventional metal finishing techniques, but their response on ABS plastics is drastically different as compared to metals. The abrasive finishing resulted in removal of unwanted material from edges and corners due to the impact of abrasion media in these mass finishing processes.

4.8.1.1   Manual sanding

Stratasys, the manufacturer of FDM machines, recommended various manual abrasive methods using sandpaper (120 and 320 grit), steel wool and filing to remove small bits of supports, burrs and hair-like strands. A hot knife can remove seams and be used for filling gaps with raw material manually.

Being simple and economical, the manual methods are not uniform, measured, consistent and precise as these methods rely upon the skill of the operator. Moreover, when dealing with parts having intricate shapes, manually finishing methods lack fidelity to the application of rapid tooling and casting.

4.8.1.2   Abrasive milling

The abrasive machining action of sandpapers of different grit sizes rotated on a wheel can be used as a material removal process through chip formation. Faster feed, speeds and smaller media increased the material removal rate, which also enhanced the surface hardness. The abrasive milling using bulk lamellar abrasive paper can achieve up to 90% improvement in surface finish with very little deviation in dimensions. In recent studies, the abrasive milling followed by physical vapour deposition showed little dimensional variations as compared to micro sandblasting and electroplating. The CNC milling using a program according to part geometry using adaptive slicing can also be used, but this subtractive machining method is unable to access the features of parts having intricate shapes.

4.8.1.3 Abrasive flow machining

Abrasive flow machining (abrasive blasting) is a process in which an abrasive laden elastic media is used for finishing and polishing the parts. In this method, a high-velocity jet of abrasives impinges on a rough surface and burrs until they are smoothened. It was found that media pressure, grit size and blasting time played a major role in controlling the final surface roughness while small dimensional changes occurred due to material removal. Abrasive flow machining has been conventionally used for finishing and sealing conformal channels for profile edge laminae tooling applications. Leong et al. [22] reported 70% improvement in surface roughness of plastic parts through abrasive jet deburring using dry air as carrier and glass beads as abrasive media. Uneven material removal is a major disadvantage due to the inability to control pressure in media, which caused weight loss and up to a 5.85% reduction in thickness. The flat surfaces showed the highest response while the process is very random and aggressive as it damaged edges and corners.

4.8.1.4 Sandblasting

Sandblasting has also been implemented by several researchers and improved surface roughness up to 96%. It has been recommended by Stratasys to be used after vapour smoothing and proved to be an ultra-fine finishing process. It is capable of giving a matte finish to parts because vapour and chemical exposure emanated the glossy surface.

4.8.1.5 Vibratory bowl finishing

In a vibratory bowl finishing, mass finishing of parts is done by abrasive action media and fine abrasive compounds mixed with water through a vibrating spring-mass system. There are two opposing sets of eccentric weights attached at the end of a shaft driven by a belt. Different material removal rates are achieved depending upon time, media size, media shape, media weight and compounds. There is a higher role of abrasive media shape and type in the surface finishing process, as the complex-shaped media get jammed in the parts. The finishing of ABS parts through vibratory bowl could enhance both surface roughness and hardness using pyramid media shape for longer machining times (3–4 h). It is therefore suggested to continue machining for longer hours using a lower media weight, which would yield better dimensional stability for ABS parts.

4.8.1.6   Barrel tumbling

Barrel tumbling is a mass finishing process used for deburring, fine finishing and conditioning of surface morphology of parts. The parts are loaded in a closed rotating tube with an abrasive compound, media and water. The process requires low initial, running and maintenance costs and is capable to machine different geometries without using any fixtures. The mathematical models can be formulated based on time, orientation angle and layer of thickness of FDM parts to predict material removal rate in the barrel finishing process [23]. Researchers used ceramic media of different geometries and reported a 52% decrease in surface roughness for triangular media. Rotation speed, machining time and media shape and size came out to be major parameters controlling the material removal rate. The orientation angles of FDM parts and barrel finishing time are the most significant parameters in the case of surface roughness and dimensional accuracy. The maximum material removal rate occurs at a 90° orientation angle while minimum at 18°.

4.8.1.7   Hot cutter machining

The cutter with a straight heated edge was designed to machine the flat plastic surface. Pandey et al. [24] developed an empirical model to predict the final surface roughness using adaptive slicing. The hot cutter machining proved successful for improving the surface finish where rake angle and cutting direction were major influential parameters. The virtual hybrid system has been proposed by integrating hot cutter machining with an adaptive slicing technique. The hot cutters with different shapes can be attached to a numeric controlled machine transversing in X–Y directions along the periphery of slices. Though the simulated system suggests considerable improvement in surface finish, the complex mechanism has practical fabrication complications with huge cost and time.

4.8.1.8   Ball burnishing

The ball burnishing process has been used to press the peaks of a surface profile and fill in the valleys. The workpiece was held between the head and tailstock while the burnishing tool was in the tool post of the lathe machine. The increase in depth of penetration and spindle speed enhanced the surface finish and wear of ABS parts while hardness increased with force applied by the burnishing tool.

The mechanical methods exhibit several challenges to finish the complex and intricate shapes. The abrasive action of media rounds the sharp edges and corners and distorts part geometry and dimensional stability. Moreover, deeper lying surfaces within grooves, notches, or other such indentations are more difficult for the abrasives to reach. Despite the fact that mass finishing methods are effective, these are quite inconsistent to impart uniform surface finish on all the parts.

4.8.2 Chemical finishing techniques

The response of ABS plastics to the chemicals, vapours and coatings has been discussed under this section as plastics have a penetrable surface, which could be easily modified by chemical action. The major precedence of chemical finishing over mechanical finishing techniques is that there is no contact of tools with the surface, which could ensure better dimensional and geometrical stabilities.

4.8.2.1 Manual painting

Few chemicals such as acetone, thinner, heptane and toluol smooth the surface of the ABS parts by chemically dissolving the rough edges and bonding the layers together. Stratasys suggested the use of several paints for cleaning and polishing of ABS parts. Although manual application of these chemical and paints saves time and cost, manual methods are non-uniform and thus deliver an uneven surface finish.

4.8.2.2 Acetone dipping

Acetone, being a primary component of cleaning agents, has good surface peeling characteristics and has been extensively used for polishing and finishing the products. The FDM parts are dipped in acetone solution (90% dimethyl ketone and 10% water) for a specific time for surface finishing. The authors also used acetone dipping as a post-processing technique where parts were dipped in an acetone bath for 5 min to get the desired surface finish. The immersion of parts in acetone solution also enhanced flexural strength, water tightness, humidity and wear resistance but slightly reduced the tensile strength. The increase in weight, ductility, flexural strength and compressive strength has been noticed after dipping in acetone while 1% shrinkage was recorded. The authors extended the study to optimize the process parameters – that is, solution concentration, time, initial roughness and temperature of chemicals using DOE and ANOVA. The results exhibited that the concentration, concentration-temperature interaction and concentration-time interaction have the highest impact on surface roughness. The acetone dipping can also be utilized for sealing and surface finishing of the ABS parts for biomedical micro device applications. The optimum conditions were exposure of 1–8 h with a 60% aqueous solution for best surface finish (up to nanoscale) and preserving part features. The use of FDM parts can be extended to develop sealants involving fluid pressure applications by decreasing porosity of ABS parts by chemical treatment. The chemically treated FDM parts are capable to withstand fluid pressure up to pressure of 276 KPa with minute dimensional changes. Moreover, parts that underwent chemical treatment showed less dimensional changes as compared to mechanically brushed parts. The advancements in experimental setup can be made with provision of a fan for forced circulation of acetone vapours in a closed chamber. The surface finish is also enhanced with an increase in rpm of a fan whereas longer exposure is required to finish ABS parts with a larger surface. Even the cold acetone vapours (at 20°C in closed container) can also be used for finishing as the rapid vapour-plastic interaction can be controlled by reducing temperature. The part surface quality is improved with minimum changes in dimensions up to an exposure of 40 min. However, at longer durations (90 min), the corners and sharp edges are rounded off [15].

The post-processing with acetone proved to be a highly effective, easy, fast and economic surface finishing process. But there is a risk of eroding and dissolving small features of parts for longer durations, and using undiluted acetone while use of a dilute solution can extend the immersion time. Still, the method needs to be controlled, automated and mechanized to carry out more accurate and systematic finishing.

4.8.2.3   Electroplating

The FDM parts are etched to clean the surface and then coated with palladium, which acts as a bonding agent. Finally, the chromium is deposited, which provides required durability. The methodology can be adopted to coat the wind tunnel models to reduce cost and fabrication times. Models used without surface finishing do not produce satisfactory results because the highly smooth surface is required for high-speed testing. A thin layer of chromium electroplated on an FDM model enhanced the surface finish, wear resistance and mechanical strength. The aerodynamic testing results showed a chromium-coated model as having a good lift capability as compared to an uncoated model. The copper coating on ABS parts also showed improvement in hardness, corrosion resistance and tensile strength. The coatings of copper, nickel and chrome tend to increase the impact strength and hardness in addition to surface finish with an increase in coating layer thickness. The major limitation of coating and electroplating is the added cost of the process and limited field of applications. The method induces mechanical strength and hardness, but dimensions of parts are also altered due to the filming of the outer surface. Moreover, metal coatings on ABS replicas may augment practical complications during investment castings for rapid tooling applications.

4.8.2.4   Vapour smoothing

An advanced finishing technique has been developed to ensure uniform heating in a chemically controlled environment for surface finishing of ABS parts. Initially, parts are allowed to cool for a few minutes in a drying chamber (pre-cooling) and then placed in a smoothing chamber (smoothing) for 10–30 s as shown in Figure 4.12. Again, the parts are placed in a cooling chamber (post-cooling), which completes one cycle. The whole cycle can be repeated multiple times till the desired surface finish is

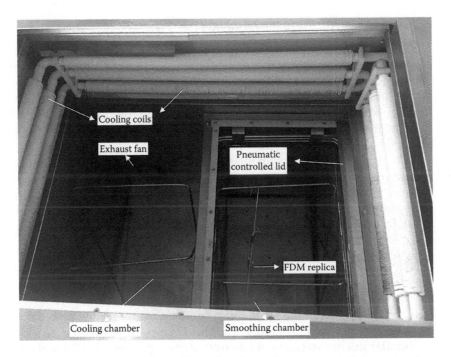

Figure 4.12 Cooling and smoothing chambers of vapour smoothing apparatus. (From Chohan, J.S. and Singh, R., *Prog. Addit. Manuf.*, 1, 105–113, 2016 [25].)

achieved. The fixed amount of solvent is heated in a smoothing chamber, and thus (chemical) vapours rise to get deposited on the ABS part, which is hung by suitable arrangement inside. Vaporized solvent fumes rise to get condensed on the ABS part, penetrate through the surface and temporarily flatten the surface because of surface tension. Initial and final readings depicted negligible dimensions changes with an extremely smooth surface [15].

4.9   Effect of vapour smoothing process on surface properties

The acetone abrades the upper surface, corners and affects the tensile strength of ABS parts. On the other hand, the vapour smoothing technique is somehow eminent to enhance the surface finish of parts and could be enabled for mass finishing. The effect of the vapour smoothing process on surface roughness, surface hardness and dimensional accuracy is discussed in the next sections.

4.9.1   Surface finishing phenomenon

The geometrical representation of the surface roughness profile has been drawn, which shows the semicircular profile of the upper surface obtained after fabrication through FDM – that is, before vapour smoothing (Figure 4.13). The nozzle extrudes a semi-molten layer of ABS, which slightly expands after deposition and thus increases the road width. Initially, there is a considerable difference between mean line and peak height (H_0) – that is, before smoothing.

The layers of ABS thermoplastic undergo localized viscous mass transport as semi-molten plastic flows from peaks into the valleys after vapour smoothing. After cooling, the viscous material settles as smooth surface under the effect of surface tension. The surface tension forces tend to attain minimum surface area, which is only possible with a smooth surface – that is, the absence of peaks and valleys. This results in a decrease in peak height by following the relation as:

$$H_0 > H_1 > H_2 > H_3$$

Here, H_1, H_2 and H_3 are the difference between the mean line and maximum height after the first, second and third cycles, respectively.

Similar inferences can be made from surface profiles of FDM replicas plotted by a surface roughness tester. After vapour smoothing, the significant decrease in maximum height as well as average surface roughness can be noticed (Figure 4.14). The increase in the number of cycles and

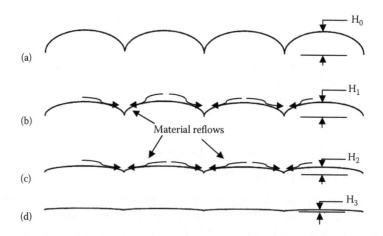

Figure 4.13 Effect of number of cycles on profile height (a) before smoothing; (b) after first cycle; (c) after second cycle and (d) after third cycle measurement. (From Chohan, J.S. et al., *Measurement*, 94, 602–613, 2016 [26].)

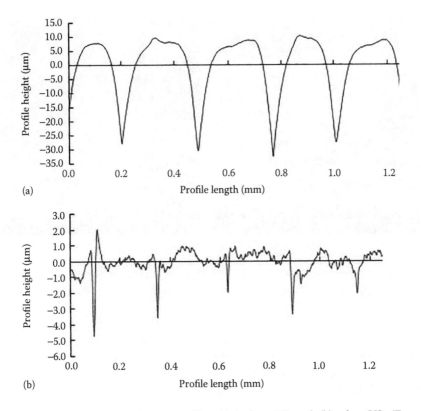

(a)

(b)

Figure 4.14 Surface roughness profiles (a) before VS and (b) after VS. (From Chohan, J.S. et al., *Measurement*, 94, 602–613, 2016 [26].)

smoothing time has a positive and direct impact on the surface finish of ABS replicas [26].

The upper and transverse view of the surface has been viewed under a scanning electron microscope (SEM) at various resolutions. The SEM micrographs images have been procured to support the smoothing phenomenon.

The minute traces of plastic material that protrude out at the intersection of adjacent layers get melted and settle down in the valley. The transverse view of the upper surface of the FDM sample depicts the semicircular profile (Figure 4.15) as observed in roughness profiles. The thin circular layers are deposited by an FDM nozzle, which creates noticeable roughness on the surface when measured perpendicular to lay direction. The SEM image of the FDM replica exposed for three cycles and 20 s smoothing time have been procured. The transverse view of the finished sample shows an extremely smooth surface when viewed at 100 x magnification (Figure 4.15). The surface becomes highly

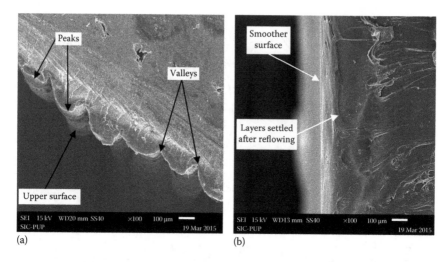

(a)   (b)

Figure 4.15 SEM micrograph (a) before vapour smoothing and (b) after vapour smoothing. (From Chohan, J.S. et al., *Measurement*, 94, 602–613, 2016 [26].)

reflective and shiny due to the absence of surface abnormalities after the vapour smoothing process. The smooth and glossy finish on the upper surface of the FDM parts was observed after visual inspection as reported by many researchers [27,28].

4.9.2   Surface hardening phenomenon

The percentage change in surface hardness is directly proportional to the post-cooling time, and next influential parameter is orientation angle. Although the part density affects the initial hardness, it does not affect the final hardness after the vapour smoothing process. The vapours do not enter deep inside the surface, but they are absorbed by the uppermost layers of the surface. The importance of post-cooling has been highlighted after analysis of surface hardness of ABS replicas. Although the post-cooling time was insignificant for surface roughness, it has maximum influence on surface hardness. The cooling chamber is maintained at 0° temperature as refrigerant continuously circulates through the cooling coils. The cooling chamber also facilitates the exhaust of vapour fumes on the surface of replicas through a fan located at the bottom section as shown in Figure 4.16. After exposure to hot vapours in the smoothing chamber, the upper layer becomes soft due to a partial meltdown. Thus, immediate cooling is required for resettlement and hardening of the surface. As the FDM parts are cooled for longer duration, the surface becomes both smoother and harder [26].

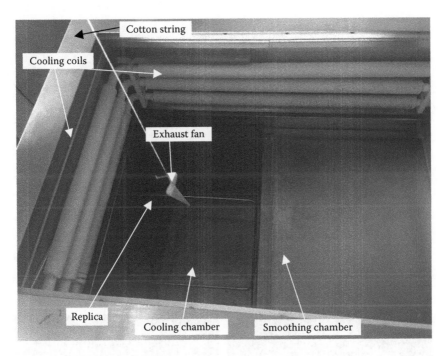

Figure 4.16 Post-cooling of replica in progress. (From Chohan, J.S. et al., *J. Manuf. Process.*, 24, 161–169, 2016 [29].)

The overall impact of the vapour smoothing process is to increase the surface hardness of FDM parts. The hardness is measured as the capability of the surface to retard the indentation. Initially (before smoothing), the surface is rough with a larger roughness profile, which is easier to penetrate as visualised in Figure 4.17a. The indenter enters more easily as surface layers are uneven and can be easily pierced. It was observed that replicas fabricated at a 90° orientation angle have lower values of surface hardness before smoothing as compared to a 0° orientation angle. This phenomenon can be explained by the fact that replicas fabricated at a 90° orientation angle have comparatively higher surface roughness. The large semicircular roughness profiles were observed in this case, which are easier to penetrate as compared to a 0° orientation angle. The material reflows from peaks down to valleys, and empty spaces are filled with plastic material after vapour exposure. After the VS process, the smooth roughness profiles are achieved, which efficiently restrict the piercing by the indenter (Figure 4.17b).

The reason can be explained by the same phenomenon of material transport. The higher material transport occurs for parts having larger roughness profiles as viewed in SEM images (Figure 4.18). The higher

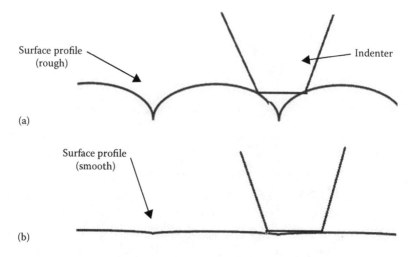

Figure 4.17 Resistance to penetration for (a) rough surface and (b) smooth surface.

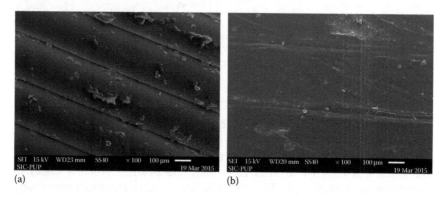

Figure 4.18 SEM images of FDM replica (a) before VS and (b) after VS. (From Chohan, J.S. et al., *J. Manuf. Process.*, 24, 161–169, 2016 [29].)

values of roughness resulted in a larger concentration of plastic material at the surface after freezing. Also, the micro-density of the surface increases after vapour smoothing as plastic fills the void spaces. After exposing the parts in a cooling chamber, the surface becomes harder due to freezing of melted plastic. The increase in post-cooling time further enhances the surface hardness due to rapid cooling. In fact, the post-cooling time has maximum impact on final hardness of parts because complete exhaust of chemical vapour fumes from the upper surface of parts is ensured after longer post-cooling duration.

4.9.3   Shrinkage phenomenon

The ABS thermoplastic undergoes shrinkage due to rapid cooling after extrusion from the nozzle. The shrinkage is automatically compensated by FDM software depending upon dimension and shape. Thus, the FDM generally fabricates oversized dimensions to eliminate the shrinkage error after cooling. In spite of the inclusion of a shrinkage compensation factor, small deviations have been experienced in complex design features of FDM parts. The layers of heated ABS material are extruded, which generate internal stresses during cooling. This causes deformation and also adds to the shrinkage factor. The radial and circular features are generally found to be manufactured undersized due to approximation error by most of the commercially available FDM machines. On the other hand, in the case of linear dimensions with intricate design, the oversized parts have been manufactured.

After vapour smoothing, the ABS plastic material flows down and gets settled in the valleys. This reduces the surface roughness as molten plastic has a tendency to cover minimum surface area, which is only possible in the case of a plain surface. The similar impact has been observed in dimensional accuracy of replicas after vapour smoothing. The hot chemical vapours lower the glass transition temperature of ABS plastic, which causes temporary and localized melting of upper layers. The peaks are lowered due to material transport, which imparts shrinkage in FDM replicas. Thus, the accuracy of linear dimensions increases, which are originally oversized. After vapour smoothing, the shrinkage of parts resulted in improvement in dimensional stability of linear sections. On the other hand, the dimensional accuracy of radial sections is negatively affected after vapour smoothing. The smoothing phenomenon is sketched in Figure 4.19, which also shows the deviation between original and actual surfaces. Figure 4.19a shows deviation in linear section having oversized dimension. After vapour smoothing, the layers reflow and settle as a smooth surface, which reduces the deviation (Figure 4.19b). On the other hand, the radial dimension is produced undersized (Figure 4.19c). After smoothing, the surface roughness is reduced but deviation increases in the radial section (Figure 4.19d). The parts must be immediately cooled after smoothing to avoid overheating, which signifies the importance of post-cooling.

The SEM micrographs can be used to correlate the layer shrinkage phenomenon with actual images (Figure 4.15). The side view of a semicircular profile of surface roughness is clearly visible before vapour smoothing while a smooth layer is observed after vapour smoothing. The semi-molten plastic material fills in the empty spaces in valleys, which reduces the part dimensions in addition to finishing.

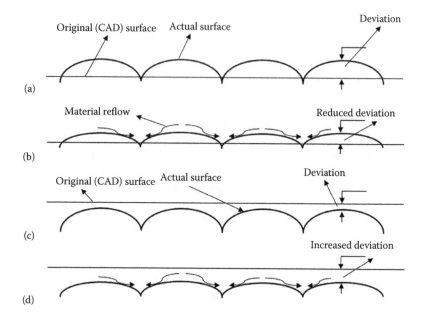

Figure 4.19 Deviations in (a) linear dimension before VS; (b) linear dimension after VS; (c) radial dimension before VS and (d) radial dimension after VS.

4.9.4 Reduction in stair-stepping

The stair-stepping phenomenon can be explained by the fundamental manufacturing principle of FDM as discussed in previous sections. During fabrication of slant sections, the extrusion nozzle has to be raised periodically, which generates a stair-like profile. The stair-stepping on the stem section of a hip replica can be observed in Figure 4.20. The stair-stepping deteriorates both the surface finish and dimensional accuracy of FDM parts.

The comparison between original CAD dimension and actual surface with stair-stepping has been visualized in Figure 4.21. After fabrication under FDM (before vapour smoothing), the significant difference between CAD and actual surface is visible. As the smoothing time is increased, the material reflowing rate increases, which consequently decreases the height of stair step.

The large stair-step is visible in Figure 4.21a, which depicts the condition of the surface before smoothing. After smoothing for 10 s, the material transport occurs due to a partial meltdown, which reduces the height of the stair-step (Figure 4.21b). The replicas are immediately hung (post-cooled) in a cooling chamber after smoothing, which results in immediate freezing of plastic material. This reduces the height of the stair-step and also the part dimensions, which explains the reason for shrinkage in replicas after vapour smoothing. The condition of the stair-step after smoothing

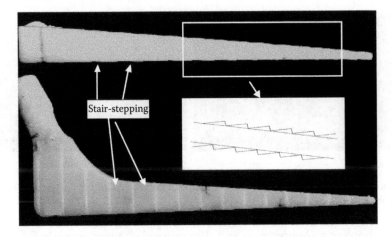

Figure 4.20 Stair-stepping visible on stem section of replica observed after fabrication. (From Chohan, J.S. and Singh, R., *Prog. Addit. Manuf.*, 1, 105–113, 2016 [25].)

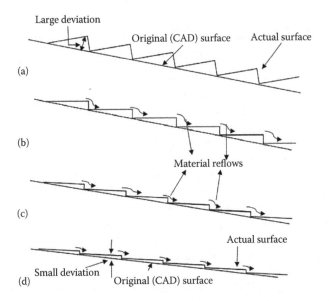

Figure 4.21 Stair-stepping (a) before smoothing (b) after smoothing for 10 seconds (c) after smoothing for 15 seconds (d) after smoothing for 20 seconds. (From Chohan, J.S. and Singh, R., *Prog. Addit. Manuf.*, 1, 105–113, 2016 [25].)

for 15 and 20 s has been shown in Figure 4.21c and d respectively. The increase in smoothing time further results in a greater reduction in amplitude of stair-step. After vapour smoothing for 20 seconds, the minute stair-step is viewed, which validates the phenomenon explained earlier. The SEM micrographs of the stair-step have been acquired to validate

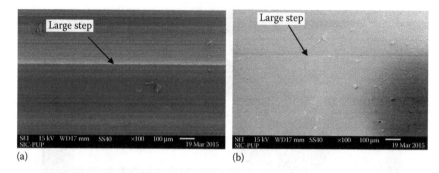

Figure 4.22 SEM micrograph of stair-step (a) before VS and (b) after VS. (From Chohan, J.S. and Singh, R., *Prog. Addit. Manuf.*, 1, 105–113, 2016 [25].)

the smoothing phenomenon and dimensional changes occurring during the finishing process (Figure 4.22). The difference between the magnitude of the stair-step before and after smoothing can be clearly differentiated.

4.10   Mathematical modelling using Buckingham Pi theorem

The Buckingham Pi theorem of dimensional analysis has been applied on FDM and VS processes for mathematical modelling. The dimensional analysis is an effective tool for generating the dimensionally homogenous and analytical equations with a large number of variables. It uses basic dimensions of variables of any complex physical system to form a simple relationship and works on the principle of equivalence between any two different entities. The theorem states that if any physical problem consists of 'n' number of variables and 'm' number of primary dimensions, it can be reduced to a single relationship having 'n − m' number of independent dimensionless variables to assume. The equations derived from the assumed independent variables can be further used for formulating dimensional relationships [30]. The following assumptions are made for model building:

1. Out of many FDM process-related parameters, only two important parameters are considered.
2. There exists a non-linear relation among parameters themselves and with responses.
3. The analysis of results is performed at a 95% of confidence level.
4. The standard specimen and test conditions are used.

The generalized mathematical model for prediction of each response has been generated using various process parameters of FDM and VS processes using the Buckingham Pi theorem. After generating a generalized

Table 4.2 Significant parameters for each response

Response	Significant parameters	Relation with response
Surface roughness	Smoothing time	Inverse
	Number of cycles	Inverse
Surface hardness	Post-cooling duration	Direct
	Orientation angle	Direct
Dimensional accuracy of radial features	Smoothing time	Inverse
	Orientation angle	Direct
Dimensional accuracy of linear features	Smoothing time	Direct
	Orientation angle	Direct

Source: Chohan, J.S. et al., *Measurement*, 94, 602–613, 2016; Chohan, J.S. et al., *Compos. Part B Eng.*, 117, 138–149, 2017 [26,31].

equation, further experiments were conducted by varying significant parameters (Table 4.2) to complete the mathematical equation. The mathematical equations could be used by industries to predict surface roughness of FDM parts before the production phase, which may reduce the manufacturing time and cost.

4.10.1   Mathematical modelling surface roughness

The orientation angle (α) does not have fundamental dimensions (MLT), and thus it cannot be considered as a valid parameter. But the initial surface roughness is strongly affected by the orientation angle, which makes it an important parameter. It is assumed that the initial surface roughness (R_{ai}) depends upon the orientation angle, which gives the following relation:

$$R_{ai} = f(\alpha) \tag{4.1}$$

Similarly, number of cycles (N), being dimensionless, is also excluded from the initial assumptions of parameters, but it has been considered in later stages because it is the second most significant parameter. Besides initial surface roughness, the other input parameters are density, pre-cooling time, smoothing time and post-cooling time. Although the FDM apparatus identifies the density of parts as low, high and solid, the basic units of density ($M^1 L^{-3} T^0$) – that is, weight/volume – are considered. The density of FDM replicas can be calculated by dividing the average weight of part by its volume. In a case study, the average value of density of replicas was found to be 0.61477 g/cm^3 for low density, 0.82092 g/cm^3 for high density and 0.94578 g/cm^3 for solid filling settings. The output parameter has been assumed as a change in surface roughness in place of percentage change because the output parameter also must have valid primary dimensions. The change in surface roughness is calculated as the difference between the initial surface roughness (R_{ai}) and final surface roughness (R_{af}) and is given as:

$$\Delta R_a = R_{ai} - R_{af} \tag{4.2}$$

Now, the output parameter has valid fundamental units ($M^0 L^1 T^0$).

As the basic dimensions under assumption are as Mass (M), Length (L) and Time (T), the following input parameters, their units and dimensions have been assumed to generate a dimensional model:

1. Initial roughness (R_{ai}) in µm $M^0 L^1 T^0$
2. Part density (ρ) in g/mm³ $M^1 L^{-3} T^0$
3. Pre-cooling time (T_{PC}) in minutes $M^0 L^0 T^1$
4. Smoothing time (T_S) in seconds $M^0 L^0 T^1$
5. Post-cooling time (T_{PT}) in minutes $M^0 L^0 T^1$

It is assumed that output depends upon all input parameters. This can be depicted as:

$$\Delta R_a = f\left(R_{ai}, \rho, T_{PC}, T_S, T_{PT}\right) \tag{4.3}$$

As per standard procedure of dimensional modelling, the values of n and m are calculated to generate Π terms. In present context:

$$n = 6 \text{ and } m = 3$$

There is a requirement of three (n − m = 3) independent dimensionless Π terms – that is, Π_1, Π_2 and Π_3. The three repeating variables – that is, initial roughness, density and post-cooling time – go directly in Π group while the other three appear once in each Π term. The selection of repeating variables has been made as suggested by Buckingham Pi theorem where each parameter represents at least one dimension – that is, mass, length and time. The Π terms can be written as:

$$\Pi_1 = \Delta R_a \left(R_{ai}\right)^{a1} \left(\rho\right)^{b1} \left(T_{PT}\right)^{c1} \tag{4.4}$$

$$\Pi_2 = T_{PC} \left(R_{ai}\right)^{a2} \left(\rho\right)^{b2} \left(T_{PT}\right)^{c2} \tag{4.5}$$

$$\Pi_3 = T_S \left(R_{ai}\right)^{a3} \left(\rho\right)^{b3} \left(T_{PT}\right)^{c3} \tag{4.6}$$

As the Π terms are dimensionless, the dimensions are substituted in Equations 4.4 through 4.6 and equated to zero to achieve the ultimate exponent of each basic dimension. Thus a^i, b^i and c^i can be solved where $i = 1, 2$ and 3. Solving for Π_1:

$$\Pi_1 = (L)(L)^{a_1}\left(ML^{-3}\right)^{b_1}(T)^{c_1}$$

Equating terms in L.H.S equal to $M^0 L^0 T^0$

$$b_1 = 0, c_1 = 0, b_1 + a_1 - 3b_1 = 0 \text{ and thus } a_1 = -1$$

Thus,

$$\Pi_1 = \Delta R_a / R_{ai} \tag{4.7}$$

Similarly, solving for Equation 4.5:

$$\Pi_2 = T(L)^{a_2}\left(ML^{-3}\right)^{b_2}(T)^{c_2}$$

Equating to $M^0 L^0 T^0$ and solving we get

$$b_2 = 0, 1 + c_2 = 0; c_2 = -1, a_2 - b_2 = 0; a_2 = 0$$

Thus,

$$\Pi_2 = T_{PC} / T_{PT} \tag{4.8}$$

Similarly, we get:

$$\Pi_3 = T(L)^{a_3}\left(ML^{-3}\right)^{b_3}(T)^{c_3}$$

Equating to $M^0 L^0 T^0$ and solving we get similar results as previous

$$b_3 = 0, 1 + c_3 = 0; c_3 = -1, a_3 - b_3 = 0; a_3 = 0$$

Thus,

$$\Pi_3 = T_S / T_{PT} \tag{4.9}$$

Thus, the ultimate relationship can be assumed as:

$$\Pi_1 = f(\Pi_2, \Pi_3)$$

Inserting values from Equations 4.7 through 4.9, we get

$$\frac{\Delta R_a}{R_{ai}} = f\left(\frac{T_{PC}}{T_{PT}}, \frac{T_S}{T_{PT}}\right)$$

Or it can be written as:

$$\frac{\Delta R_a}{R_{ai}} = \frac{K_N \cdot T_{pc}}{T_{PT}^2} \qquad (4.10)$$

where K_N is a constant of proportionality that depends upon the smoothing time (T_S) for a given number of cycles – that is, $N = 1, 2, 3$ – since these are most significant parameters.

The Equation 4.10 can be further modified to evaluate percentage change in surface roughness after the vapour finishing process.

Multiplying L.H.S and R.H.S of Equation 4.10 by 100:

$$\frac{\Delta R_a}{R_{ai}} \times 100 = \frac{100 \cdot K_N \cdot T_{pc}}{T_{PT}^2}$$

The ratio of change in roughness and initial roughness can be written as:

$$\%\Delta R_a = \frac{100 \cdot K_N \cdot T_{pc}}{T_{PT}^2} \qquad (4.11)$$

The Equation 4.11 is a generalised equation and can be used to evaluate the impact of vapour smoothing on FDM parts manufactured at any orientation angle. The equation is valid for different values of initial roughness (R_{ai}) and any part geometry as the output is in the form of percentage change ($\%\Delta R_a$).

The percentage change in surface roughness is not uniform, but it depends upon a number of cycles. There are different values of constant of proportionality (K_N) for each cycle. Also, the response depends upon the size and shape of FDM parts.

The methodology has been demonstrated to explore the values K_N by taking the example of hip implant replicas manufactured by FDM using ABS thermoplastic. Here, six replicas were manufactured at constant FDM parameters ($\alpha = 90°$, $\rho = 614.77 \times 10^{-3}\,\text{g/cm}^3$) to keep minimum manufacturing time and part/support material usage. The six replicas were exposed in a smoothing chamber for a variable smoothing time ($T_s = 5, 10, 15, 20, 25, 30\,\text{s}$). The pre-cooling time (T_{PC}) has been taken as one minute as it has no significant effect on response. In spite of being an insignificant parameter, the post-cooling time (T_{PT}) has been taken as ten minutes to ensure thorough settlement and cooling of the upper surface after smoothing. Moreover, the whole cycles are repeated thrice to evaluate the impact of a different number of cycles. The initial and final surface roughness was measured after each cycle.

The variation of average surface roughness has been plotted in Figure 4.23. The decrease in roughness values is lower for the initial 5 s

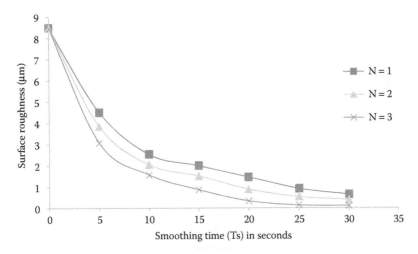

Figure 4.23 Variation of surface roughness with smoothing time for different number of cycles. (From Chohan, J.S. et al., *J. Manuf. Process.*, 24, 161–169, 2016 [29].)

while beyond 10 s smoothing time the ABS surface attains the required viscosity for reflowing. Once the plasticization stage has been reached, the parts are cooled in a cooling chamber where layers rearrange evenly under the effect of surface tension. As revealed by Figure 4.23, the final roughness is nearly the same for smoothing times of 25 and 30 s, and thus it is not recommended to run the smoothing process beyond 30 s. As the cycles are repeated, the greater reduction is achieved in average roughness while variation of roughness between consecutive cycles decreases with an increase in smoothing time and least for $T_s = 30$ s. Moreover, the smoothing time beyond 30 s may deteriorate the part surface due to overheating.

Since the Equation 4.11 delivers the result in percentage change, the values of K_N are acquired by plotting the percentage change in surface roughness with smoothing time (Figure 4.24).

The individual value of K_N has been calculated from the best-fitted curve equations for each cycle as shown in Figure 4.24. The values of constant of proportionality K_N are given as:

$$K_N = -0.071 \left(T_s^2\right) + 4.175 \, T_s + 30.38 \quad \text{if, } N = 1$$

$$K_N = -0.075 \left(T_s^2\right) + 4.164 \, T_s + 37.63 \quad \text{if, } N = 2$$

$$K_N = -0.079 \left(T_s^2\right) + 4.101 \, T_s + 46.27 \quad \text{if, } N = 3$$

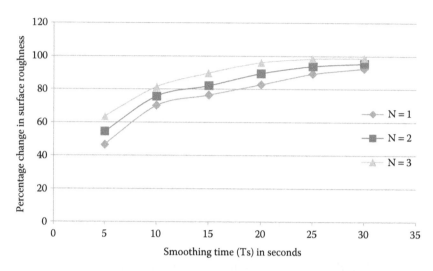

Figure 4.24 Percentage change in surface roughness with smoothing time. (From Chohan, J.S. et al., *J. Manuf. Process.*, 24, 161–169, 2016 [29].)

Thus, the Equation 4.11 can be rewritten as:

$$\%\Delta R_a = \frac{100.\,[-0.071\left(T_S{}^2\right)+4.175\;T_S+30.38].\,T_{pc}}{T_{PT}{}^2} \tag{4.12}$$

For one cycle (N = 1)

$$\%\Delta R_a = \frac{100.\,[-0.075\left(T_S{}^2\right)+4.164\;T_S+37.63].\,T_{pc}}{T_{PT}{}^2} \tag{4.13}$$

For two cycles (N = 2)

$$\%\Delta R_a = \frac{100.\,[-0.079\left(T_S{}^2\right)+4.101\;T_S+46.27].\,T_{pc}}{T_{PT}{}^2} \tag{4.14}$$

For three cycles (N = 3)

The mathematical equations have been formulated. Finally, the validity of equations was confirmed by Figure 4.25, which compared predicted data with experimental data. In general, the proposed mathematical model is capable to predict the response. Thus, it can be concluded that the mathematical model formulated using the Buckingham Pi theorem can accurately predict surface roughness of ABS replicas finished by the vapour smoothing process.

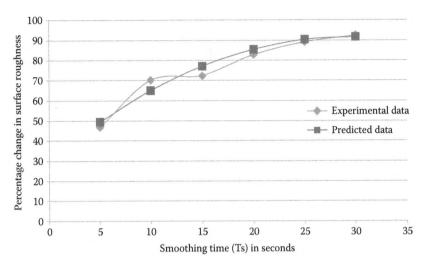

Figure 4.25 Comparison of modelled (predicted) data and experimental data. (From Chohan, J.S. et al., *J. Manuf. Process.*, 24, 161–169, 2016 [29].)

4.10.2   Mathematical modelling of surface hardness

The durometer measures the shore D hardness in durometer units, which are converted into valid units of force before calculations. The hardness is assumed as force applied on an indenter, which is calibrated with deflection in spring attached to it. Thus, the spring force can be calculated from the hardness value as per ASTM D2240 standards, and the relation between spring force and durometer units is given as:

$$1\,H_d = 0.4445\,\text{Newton}\,(N) \tag{4.15}$$

As an example, the hardness value of 50 units is assumed as 22.225 N. Thus, the unit of hardness is assumed as Newton ($M^1\,L^1\,T^{-2}$).

The other input parameters having valid units are included in the calculations. Since the orientation angle has significant contribution, it must be included in the dimensional model, but it does not have valid units. The orientation angle has significant impact on initial surface roughness (R_{ai}), which in turn affects the surface hardness (H_d) as discussed in the previous section. Thus, the following relation can be assumed as:

$$H_d\,\alpha\,\text{Orientation angle} \tag{4.16}$$

Also,

$$R_{ai} \ \alpha \ \text{Orientation angle} \qquad (4.17)$$

Thus, from relation 5.16 and 5.17, it can be written that:

$$H_d \ \alpha \ R_{ai} \qquad (4.18)$$

Thus, the initial surface roughness is included as an input parameter as a substitute to orientation since both have similar impact on hardness as discussed earlier. On the other hand, a number of cycles cannot be included in the model as it neither has valid units nor is it a significant parameter for hardness.

Thus, the following input parameters have been assumed to generate a dimensional model:

1. Initial surface roughness R_{ai} in $\mu m \ M^0 L^1 T^0$
2. Density (ρ) in $g/mm^3 \ M^1 L^{-3} T^0$
3. Pre-cooling time (T_{PC}) in minutes $M^0 L^0 T^1$
4. Smoothing time (T_S) in seconds $M^0 L^0 T^1$
5. Post-cooling time (T_{PT}) in minutes $M^0 L^0 T^1$

The hardness depends upon these five parameters, and the following relation can be written as:

$$H_d = f\left(R_{ai}, \rho, T_{PC}, T_S, T_{PT}\right) \qquad (4.19)$$

Since there are six parameters, the value of 'n' is 6 and 'm' is 3. So there are three (n − m = 3) independent dimensionless Π terms – that is, Π_1, Π_2 and Π_3. The three repeating variables – that is, initial roughness, density and smoothing time – are included directly into Π group while the other three appear once in each Π term.

The Π terms can be written as:

$$\Pi_1 = H_d \left(R_{ai}\right)^{a_1} \left(\rho\right)^{b_1} \left(T_S\right)^{c_1} \qquad (4.20)$$

$$\Pi_2 = T_{PC} \left(R_{ai}\right)^{a_2} \left(\rho\right)^{b_2} \left(T_S\right)^{c_2} \qquad (4.21)$$

$$\Pi_3 = T_{PT} \left(R_{ai}\right)^{a_3} \left(\rho\right)^{b_3} \left(T_S\right)^{c_3} \qquad (4.22)$$

As the Π terms are dimensionless, the dimensions are substituted in Equations 4.20 through 4.22 and equated to zero to achieve the ultimate exponent of each basic dimension. Thus a^i, b^i and c^i can be solved where $i = 1, 2$ and 3. Solving for Π_1:

$$\Pi_1 = \left(MLT^{-2}\right)\left(L\right)^{a_1}\left(ML^{-3}\right)^{b_1}\left(T\right)^{c_1}$$

Equating terms in L.H.S equal to $M^0 L^0 T^0$

$$b_1 = -1, c_1 = 2, -3b_1 + a_1 + 1 = 0 \text{ and thus } a_1 = -4$$

Thus,

$$\Pi_1 = H_d \left(R_{ai}\right)^{-4}\left(\rho\right)^{-1}\left(T_S\right)^{2} \tag{4.23}$$

Similarly, solving for Equation 4.21:

$$\Pi_2 = T\left(L\right)^{a_2}\left(ML^{-3}\right)^{b_2}\left(T\right)^{c_2}$$

Equating to $M^0 L^0 T^0$ and solving we get

$$b_2 = 0, 1 + c_2 = 0; c_2 = -1, a_2 - b_2 = 0; a_2 = 0$$

Thus,

$$\Pi_2 = T_{PC}/T_S \tag{4.24}$$

Similarly, we get:

$$\Pi_3 = T\left(L\right)^{a_3}\left(ML^{-3}\right)^{b_3}\left(T\right)^{c_3}$$

Equating to $M^0 L^0 T^0$ and solving we get similar results as previous

$$b_3 = 0, 1 + c_3 = 0; c_3 = -1, a_3 - b_3 = 0; a_3 = 0$$

Thus,

$$\Pi_3 = T_{PT}/T_S \tag{4.25}$$

The ultimate relationship between Π terms can be written as:

$$\Pi_1 = f\left(\Pi_2, \Pi_3\right) \tag{4.26}$$

Inserting values from Equations 4.23 through 4.25 into 4.26, we get

$$\frac{H_d\left(Ts\right)^2}{\rho\left(R_{ai}\right)^2} = f\left(\frac{T_{PC}.T_{PT}}{Ts.Ts}\right)$$

$$\frac{H_d\left(Ts\right)^2}{\rho\left(R_{ai}\right)^2} = \frac{K_N.T_{PC}.T_{PT}}{Ts^2}$$

$$H_d = \frac{K_N.T_{PC}.T_{PT}.\rho.\left(R_{ai}\right)^2}{Ts^4} \tag{4.27}$$

where K_N is a constant of proportionality that depends upon post-cooling time (T_{PC}).

The Equation 4.27 is a general formula that can be used to evaluate the surface hardness of FDM parts of any geometry after the vapour smoothing process.

The experimental results are shown, which were performed to complete the hardness equation by taking data from the case study of hip replicas. The seven replicas were manufactured at a 90° orientation angle and high density. The replicas were vapour smoothed with 20 min. pre-cooling time and 15 s smoothing time. The post-cooling of seven replicas was done for different durations – that is, 0, 5, 10, 15, 20, 25, 30 min – as it was the most significant parameter. The variation of hardness against the post-cooling time has been plotted in Figure 4.26.

The trend line equation of the chart has been acquired and was used to derive the value of K_N. Thus, the value of the constant of proportionality can be written as:

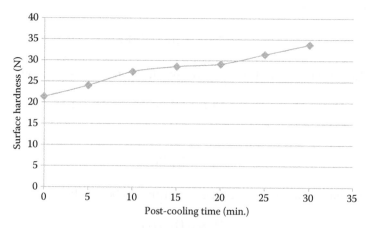

Figure 4.26 Variation of surface hardness with post-cooling time. (From Chohan, J.S. et al., *J. Manuf. Process.*, 24, 161–169, 2016 [29].)

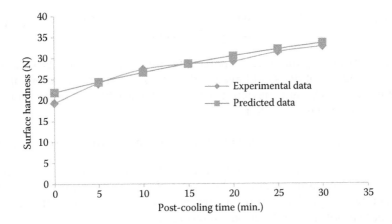

Figure 4.27 Comparison of predicted and experimental values. (From Chohan, J.S. et al., *J. Manuf. Process.*, 24, 161–169, 2016 [29].)

$$K_N = -0.004\left(T_{PC}\right)^2 + 0.505\left(T_{PC}\right) + 21.68$$

After inserting the values of K_N in Equation 5.27, the final equation can be written as:

$$H_d = \frac{\left[-0.004\left(T_{PT}\right)^2 + 0.505\left(T_{PT}\right) + 21.68\right] \cdot T_{PC} \cdot \rho \cdot \left(R_{ai}\right)^2}{Ts^4} \tag{4.28}$$

The comparison has been made between predicted and experimental values (Figure 4.27) of surface hardness, which confirm the efficacy of the mathematical model.

4.10.3   Mathematical modelling of dimensional accuracy

The dimensional accuracy is measured in percentage deviation in dimensions as:

$$\text{Percentage deviation} = \frac{\text{change in dimension}}{\text{original dimension}} \times 100 \tag{4.29}$$

The dimensions are measured before and after vapour smoothing and compared with the original CAD dimensional. Thus, the percentage deviation has been calculated twice – before vapour smoothing, which can be calculated as:

$$\%\Delta D_b = \frac{\text{original CAD dimension} - \text{actual dimension before smoothing}}{\text{original CAD dimension}} \times 100 \tag{4.30}$$

$$\%\Delta D_a = \frac{\text{original CAD dimension} - \text{actual dimension before smoothing}}{\text{original CAD dimension}} \times 100 \quad (4.31)$$

Since the output parameter is a percentage deviation, it cannot be included in the model as it does not have valid units. Thus, deviation after vapour smoothing (ΔD_a) is considered a response with fundamental units as $M^0 L^1 T^0$. It can be calculated as:

$$\Delta D_a = \text{Original (CAD) dimension} - \text{actual dimension} \quad (4.32)$$

The orientation angle has a significant contribution but could not be considered as an input parameter because it does not have valid units. The orientation angle has significant impact on percentage deviation before vapour smoothing (ΔD_b). It can be written as:

$$\Delta D_b \propto \text{Orientation angle} \quad (4.33)$$

Also, the initial deviation has direct relation with final deviation. So,

$$\Delta D_a \propto \Delta D_b \quad (4.34)$$

Thus, percentage deviation before smoothing has been considered as an input parameter. Finally, following input parameters, their units and dimensions have been assumed to generate a dimensional model:

- Initial deviation ΔD_b in mm $M^0 L^1 T^0$
- Density (ρ) in g/mm^3 $M^1 L^{-3} T^0$
- Pre-cooling time (T_{PC}) in minutes $M^0 L^0 T^1$
- Smoothing time (T_S) in seconds $M^0 L^0 T^1$
- Post-cooling time (T_{PT}) in minutes $M^0 L^0 T^1$

The percentage deviation depends upon all the aforementioned input parameters and is written as:

$$\Delta D_a \propto (\Delta D_b, \rho, T_{PC}, T_S, T_{PT}) \quad (4.35)$$

or it can be expressed as

$$\Delta D_a = f(\Delta D_b, \rho, T_{PC}, T_S, T_{PT}) \quad (4.36)$$

In the present calculation, the value of n is 6 and m is 3. So there are three (n – m = 3) independent dimensionless Π terms – Π_1, Π_2 and Π_3. The three repeating variables – initial deviation, density and pre-cooling time – will go directly in Π group while the other three appear once in each Π term.

The selection of repeating variables has been made as suggested by the Buckingham Pi theorem where each parameter represents at least one dimension – that is, mass, length and time.

$$\Pi_1 = \Delta D_a (\Delta D_b)^{a_1} (\rho)^{b_1} (T_{PC})^{c_1} \tag{4.37}$$

$$\Pi_2 = T_S (\Delta D_b)^{a_2} (\rho)^{b_2} (T_{PC})^{c_2} \tag{4.38}$$

$$\Pi_3 = T_{PT} (\Delta D_b)^{a_3} (\rho)^{b_3} (T_{PC})^{c_3} \tag{4.39}$$

As the Π terms are dimensionless, the dimensions are substituted in Equations 4.37 through 4.39 and equated to zero to achieve the ultimate exponent of each basic dimension. Thus a^i, b^i and c^i can be solved where $i = 1, 2$ and 3. Solving for Π_1:

$$\Pi_1 = \left(M^0 \, L \, T^0 \right) (L)^{a_1} \left(ML^{-3} \right)^{b_1} (T)^{c_1}$$

Equating terms in L.H.S equal to $M^0 L^0 T^0$

$$b_1 = 0, c_1 = 0, -3b_1 + a_1 + 1 = 0 \text{ and thus } a_1 = -1$$

Thus,

$$\Pi_1 = \Delta D_a (\Delta D_b)^{-1} \tag{4.40}$$

Similarly, solving for Equation 4.38:

$$\Pi_2 = M^0 \, L^0 \, T (L)^{a_2} \left(ML^{-3} \right)^{b_2} (T)^{c_2}$$

Equating to $M^0 L^0 T^0$ and solving we get:

$$b_2 = 0, 1 + c_2 = 0; \ c_2 = -1, a_2 - 3b_2 = 0; \ a_2 = 0$$

Thus,

$$\Pi_2 = T_S / T_{PC} \tag{4.41}$$

Similarly, we get:

$$\Pi_3 = T (L)^{a_3} \left(ML^{-3} \right)^{b_3} (T)^{c_3}$$

Equating to $M^0 L^0 T^0$ and solving we get similar results as previous

$$b_3 = 0, 1 + c_3 = 0; \; c_3 = -1, a_3 - b_3 = 0; \; a_3 = 0$$

Thus,

$$\Pi_3 = T_{PT}/T_{PC} \tag{4.42}$$

Thus, ultimate relationship can be assumed as:

$$\Pi_1 = f\left(\Pi_2, \Pi_3\right)$$

Inserting values from Equations 4.40 through 4.42 in an ultimate relationship, we get

$$\frac{\Delta D_a}{\Delta D_b} = f\left(\frac{T_S.T_{PT}}{T_{PC}.T_{PC}}\right)$$

$$\frac{\Delta D_a}{\Delta D_b} = \left(\frac{K_N.T_S.T_{PT}}{T_{PC}^2}\right) \tag{4.43}$$

Now, we know that

$$\Delta D_a = \text{original}\left(\text{CAD}\right) \text{dimension} - \text{actual dimension after smoothing}$$

And,

$$\Delta D_b = \text{original}\left(\text{CAD}\right) \text{dimension} - \text{actual dimension before smoothing}$$

By dividing and multiplying the Equation 4.43 by:

$$\left(\text{original CAD dimension} \times 100\right)$$

We get:

$$\frac{\Delta D_a \times \left(\text{original CAD dimension} \times 100\right)}{\Delta D_b \times \left(\text{original CAD dimension} \times 100\right)} = \left(\frac{K_N.T_S.T_{PT}}{T_{PC}^2}\right)$$

Rearranging we get:

$$\frac{\Delta D_a \times 100}{(\text{original CAD dimension})} \bigg/ \frac{\Delta D_b \times 100}{(\text{original CAD dimension})} = \left(\frac{K_N.T_S.T_{PT}}{T_{PC}^2}\right)$$

$$\frac{\%\Delta D_a}{\%\Delta D_b} = \left(\frac{K_N.T_S.T_{PT}}{T_{PC}^2}\right)$$

$$\%\Delta D_a = \left(\frac{\%\Delta D_b.K_N.T_S.T_{PT}}{T_{PC}^2}\right) \tag{4.44}$$

where K_N is a constant of proportionality.

The generalized Equation 4.44 has been achieved, which can predict percentage deviation in FDM parts after vapour smoothing. But to complete the equation value of K_N must be known. The K_N can be calculated using impact of most significant parameter which is smoothing time.

Thus, dimensional accuracy of radial and liner features of hip implant replicas has been calculated using a similar procedure as discussed in previous sections.

The values of constant of proportionality K_N are given as:

$$K_N = 0.054\,(T_s)^2 - 0.243\,(Ts) + 5.364 \quad \text{for head diameter (Radial feature)}$$

$$K_N = 0.003\,(T_s)^2 - 0.269\,(Ts) + 5.341 \quad \text{for stem thickness (Linear Feature)}$$

Thus, to predict deviation in a desired part feature, the value of K_N for respective part feature can be inserted in generalized Equation 4.44. The final equation for prediction of percentage deviation in head diameter and stem thickness is given by equation and respectively.

$$\%\Delta D_a = \left[\frac{\%\Delta D_b.(0.054\,Ts^2 - 0.243\,Ts + 5.364).T_S.T_{PT}}{T_{PC}^2}\right] \tag{4.45}$$

$$\%\Delta D_a = \left[\frac{\%\Delta D_b.(0.003\,Ts^2 - 0.269\,Ts + 5.341).T_S.T_{PT}}{T_{PC}^2}\right] \tag{4.46}$$

The predicted data was used for comparison with actual experimental data for head diameter and stem thickness, which have been shown in Figures 4.28 and 4.29, respectively.

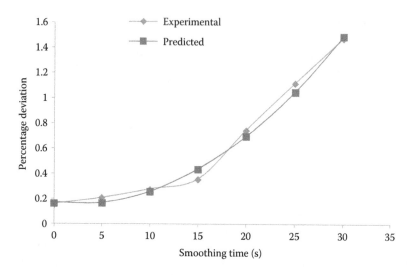

Figure 4.28 Comparison of experimental and predicted data for head diameter. (From Chohan, J.S. et al., *Compos. Part B Eng.*, 117, 138–149, 2017 [31].)

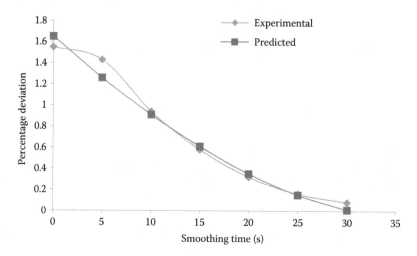

Figure 4.29 Comparison of experimental and predicted data for stem thickness. (From Chohan, J.S. et al., *Compos. Part B Eng.*, 117, 138–149, 2017 [31].)

The curves of modelled and experimental data almost superimpose with each other, which ensures the robustness of the mathematical model. The predicted and experimental values were found to be nearly the same as observed from the graph. At thirty-second smoothing time, the data points almost coincide with each other as similar in the case of head diameter.

4.11   Summary

The fused deposition modelling is a technology of the future that could revolutionise the production technology through cost-effective customized products with lower lead times. The inherent disability of poor surface finish of FDM parts could be annihilated through a vapour smoothing technique to serve a desired purpose. The nonconventional finishing technique utilizes hot chemical vapours to partially melt the upper surface of ABS parts, which yields excellent surface properties. The mathematical models formulated through the Buckingham Pi method could precisely predict the surface properties of FDM parts that have undergone vapour smoothing, which would further reduce wastage of time, material and energy.

Short answer questions

1. How is 3D printing the emerging tool for design and development?
2. Briefly describe the scope of rapid prototyping techniques in modern industry.
3. Write a short note on various types of rapid prototyping techniques.
4. Explain the working principle of FDM.
5. Justify the failure of traditional prototyping methods in modern industry.
6. Rapid tooling with FDM is possible or not. Brief it.
7. The stair-stepping effect is observed in FDM parts. How can we reduce it?
8. What is Fused Fabrication Filament?
9. What are the various factors that influence the surface quality of FDM parts?
10. Write a short note on smart materials.
11. How do the different process variables of FDM affect its quality?
12. What is the difference between rapid manufacturing and rapid tooling?
13. The pre-processing techniques utilize _____ settings of input parameters to achieve desired surface characteristics. (Answer: optimized)
14. The surface finish of FDM parts _____ with an increase in smoothing time and number of cycles during vapour smoothing process. (Answer: increases)
15. Choose the correct answer:
    i. The hardness of ABS thermoplastic is measured on the following scale:
       a. Rockwell
       b. Shore D
       c. Brinnel
       d. Shore A

   ii. The reason for poor surface finish of FDM parts is:
      a. Larger layer thickness
      b. Steep orientation angle
      c. Chordal error
      d. All of the above
  iii. Which is the mechanical finishing technique:
      a. Acetone dipping
      b. Ball burnishing
      c. Vapour smoothing
      d. Electroplating
  iv. The hot semi-molten layers of ABS plastic are smoothened under the action of
      a. Surface tension
      b. Static force
      c. Compressive force
      d. Tensile force
   v. The post-cooling of FDM part after vapour smoothing is mandatory to ensure
      a. Faster cooling of upper layers
      b. Exhaust of vapour fumes from part surface
      c. Higher hardness values
      d. All of the above

Answers:

   i. b
  ii. d
 iii. b
  iv. a
   v. d

Theoretical questions

1. Explain briefly the FDM process.
2. Classify the various types of surface finishing techniques for improving the surface quality of FDM parts.
3. Classify the various types of FDM materials.
4. Explain briefly the application area of 4D materials.
5. Write a short note on FDM in dentistry.
6. What is meant by additive manufacturing? Explain the various application areas of additive manufacturing.
7. Write a short note on application-specific materials.
8. What are the different applications of FDM technology in the medical field?

9. What are the advantages of using FDM parts as patterns for investment casting of implants?
10. What are the reasons for poor surface finish of FDM parts?
11. Why is a chordal error produced in FDM parts?
12. What are the limitations of post-processing techniques of surface enhancement?
13. Name any three post-processing techniques and explain their working principle.
14. What is the working principle of the vapour smoothing process?
15. In which conditions should vapour smoothing be preferred over conventional finishing techniques of surface finishing?
16. Why is hardness of ABS parts enhanced after the vapour smoothing process?
17. Explain the impact of the vapour smoothing technique on dimensional accuracy of FDM parts.
18. Why is Buckingham's π-theorem preferred for mathematical modelling as compared to other modelling techniques?
19. Write a short note on the importance of the vapour smoothing process in the fabrication of steel implants through investment casting.

Numerical questions (solved)

Q1. Calculate the final surface roughness of the hip implant replica finished with the vapour smoothing technique for three cycles with a smoothing time of 15 s each. Assume the replicas were fabricated at a low density with a 90° orientation angle. The pre-cooling and post-cooling durations are 1 and 10 min, respectively. Assume the initial surface roughness = 7.6540 μm.

Solution:
The following conditions are given in the statement:

$$R_{ai} = 7.6540 \, \mu m, \, \rho = 0.61477 \, g/mm^3,$$

$$T_{PC} = 1 \, min, \, T_S = 15 \, s, \, T_{PT} = 10 \, min, \, N = 3$$

The equation for percentage change in surface roughness for the three cycles is written as:

$$\%\Delta R_a = \frac{100.[-0.079(T_S^2) + 4.101 \, T_S + 46.27].T_{pc}}{T_{PT}^2}$$

Put values in the equation:

$$\%\Delta R_a = 100 \times \frac{\left[-0.079(15)^2 + 4.101(15) + 46.27\right].1}{(10)^2}$$

$$\%\Delta R_a = 100 \times \left[\frac{-17.775 + 61.515 + 46.27}{100}\right]$$

$$\%\Delta R_a = 90.01$$

Thus, there is a 90.01% decrease in surface roughness after vapour smoothing. The final surface roughness can be calculated by the following relation:

$$\%\Delta R_a = \left(R_{ai} - R_{af}/R_{ai}\right) \times 100$$

Put values: $90.01 = \left(7.6540 - R_{af}/7.6540\right) \times 100$

Rearranging: $R_{af} = 7.6540 - 6.7539 = 0.9001\ \mu m$

Answer: Final surface roughness = 0.9001 µm

Q2. Calculate the shore D hardness of ABS replicas having a medium density that underwent vapour smoothing for 5 seconds. The following data is given: Pre-cooling time = 15 min, post-cooling time = 20 min, Initial roughness = 6.9825 µm.

Solution:
The given data is written as:

$$R_{ai} = 6.9825\ \mu m,\ \rho = 0.82052\ g/mm^3,$$

$$T_{PC} = 15\ min,\ T_S = 06\ s,\ T_{PT} = 20\ min$$

The hardness is given as:

$$H_d = \frac{\left[-0.004(T_{PT})^2 + 0.505(T_{PT}) + 21.68\right].T_{PC}.\rho.(R_{ai})^2}{T_s^4}\ (\text{in Newton})$$

Put values: $H_d = \dfrac{\left[-0.004(20^2) + 0.505(20) + 21.68\right] \times 15 \times \rho \times (6.9825)^2}{(5)^4}$

$$H_d = \frac{[-1.6+10.1+21.68] \times 15 \times 0.82052 \times 48.7553}{625}$$

$$H_d = \frac{30.18 \times 15 \times 0.82052 \times 48.7553}{625} = \frac{18118.9557}{625} = 28.97$$

Thus, the final hardness comes out to be 28.99 N in terms of spring force. To convert the spring force into shore d hardness value (unitless), the following equation is used:

$$1\,H_d = 0.4445\,N$$

Answer: The hardness in terms of shore d scale can be written as 65.17.

Q3. The ABS replica of the hip joint is fabricated at a 90° orientation angle with a solid density having an original thickness of 6.6742 mm at the midsection of the stem. The replica is in a smoothing chamber. Calculate the stem thickness after finishing if the parts are vapour treated for 6 seconds and pre-cooled for 20 min and post-cooled for 10 min. Also comment on the dimensional accuracy after finishing. Assume 1.5% oversizing in fabrication by FDM.

Solution:
The following conditions are given in the statement:

$$\alpha = 90°, \rho = 0.94578\,g/mm^3, T_{PC} = 20\,min, T_S = 6\,s, T_{PT} = 10\,min,$$

$$D_o = 6.6742\,mm, \%\Delta D_b = 1.5$$

The thickness of the stem section at the middle before vapour smoothing (after fabrication) can be calculated by the relation:

$$\%\Delta D_b = \frac{D_o - D_b}{D_o} \times 100 \quad OR \quad \frac{\%\Delta D_b}{100} = 1 - \frac{D_b}{D_o}$$

$$\frac{D_b}{D_o} = 1 - \frac{\%\Delta D_b}{100} \quad OR \quad D_b = D_o \left[1 - \frac{\%\Delta D_b}{100} \right]$$

$$\text{Put values: } D_b = 6.6742 \left[1 - \frac{1.5}{100} \right]$$

$$D_b = 6.5740\,mm$$

The percentage change in dimensions after smoothing is given by the equation:

$$\%\Delta D_a = \left\{ \frac{\%\Delta D_b.\left[0.003\left(T_s\right)^2 - 0.269\left(T_s\right) + 5.341\right].T_S.T_{PT}}{T_{PC}{}^2} \right\}$$

$$\%\Delta D_a = \left\{ \frac{1.5 \times \left[0.003\left(6\right)^2 - 0.269\left(6\right) + 5.341\right] \times 6 \times 10}{20^2} \right\}$$

$$\%\Delta D_a = \left\{ \frac{1.5 \times \left[0.003\left(6\right)^2 - 0.269\left(6\right) + 5.341\right] \times 12 \times 16}{20^2} \right\}$$

$$\%\Delta D_a = \left\{ \frac{1.5 \times 3.835 \times 6 \times 10}{20^2} \right\}$$

$$\%\Delta D_a = 0.86\%$$

Answer: After vapour smoothing, the hip thickness of the stem would be 0.86% more than the CAD design. Thus, the dimensional accuracy of the product has been improved after finishing.

Q4. Calculate the change in hardness of the hip replicas that underwent vapour smoothing if the pre-cooling time is increased three times and smoothing time is doubled. Assume required values.

Solution:
The equation for hardness is given as:

$$H_d = \frac{\left[-0.004\left(T_{PT}\right)^2 + 0.505\left(T_{PT}\right) + 21.68\right].T_{PC}.\rho.\left(R_{ai}\right)^2}{T_s{}^4}$$

Case I: Let the initial surface roughness, density, pre-cooling time, smoothing time and post-cooling time be R_{ai1}, ρ_1, T_{PC1}, T_{S1} and T_{PT1} respectively.

$$H_{d1} = \frac{\left[-0.004\left(T_{PT1}\right)^2 + 0.505\left(T_{PT1}\right) + 21.68\right].T_{PC1}.\rho_1.\left(R_{ai1}\right)^2}{T_{s1}{}^4} \tag{4.47}$$

Case II: Ff pre-cooling time is increased three times and smoothing time is doubled:

$$H_{d2} = \frac{\left[-0.004\left(T_{PT2}\right)^2 + 0.505\left(T_{PT2}\right) + 2.68\right].T_{PC2}.\rho_2.\left(R_{ai2}\right)^2}{T_{s2}^{\,4}} \quad (4.48)$$

Dividing Equation 4.48 by 4.47:

$$H_{d1}/H_{d2} = \frac{\dfrac{\left[-0.004\left(T_{PT1}\right)^2 + 0.505\left(T_{PT1}\right) + 21.68\right].T_{PC1}.\rho_1.\left(R_{ai1}\right)^2}{T_{s1}^{\,4}}}{\dfrac{\left[-0.004\left(T_{PT2}\right)^2 + 0.505\left(T_{PT2}\right) + 21.68\right].T_{PC2}.\rho_2.\left(R_{ai2}\right)^2}{T_{s2}^{\,4}}}$$

Here, $R_{ai1} = R_{ai2}$ and $\rho_1 = \rho_2$. Assume the post-cooling time (T_{PT}) as 15 min

$$H_{d1}/H_{d2} = \frac{\dfrac{\left[-0.004\left(15\right)^2 + 0.505\left(15\right) + 21.68\right].T_{PC1}}{T_{s1}^{\,4}}}{\dfrac{\left[-0.004\left(15\right)^2 + 0.505\left(15\right) + 21.68\right].T_{PC2}}{T_{s2}^{\,4}}}$$

$$H_{d1}/H_{d2} = \frac{T_{PC1}}{T_{s1}^{\,4}} \Big/ \frac{T_{PC2}}{T_{s2}^{\,4}}$$

As per statement, $T_{PC1} = T_{pc2}/3$ and $T_{s1} = T_{s2}/2$. Put values:

$$H_{d1}/H_{d2} = \frac{3.T_{PC2}}{16 \times T_{s2}^{\,4}} \Big/ \frac{T_{PC2}}{T_{s2}^{\,4}}$$

$$H_{d1}/H_{d2} = 3/16 = 0.1875$$

$$H_{d2} = 5.33\left(H_{d1}\right) \quad \text{OR} \quad H_{d2} = \left(1 + 4.33\right)H_{d1}$$

Thus, hardness in case II would be 433% higher than case I.

Numerical questions (unsolved)

Q1. The hip implant is vapour smoothed for two cycles with 10 seconds each and later cooled in a cooling chamber for 15 min. The replicas are fabricated at high density with a 0° orientation angle. Calculate the percentage improvement in surface finish and final roughness of the hip implant replica if the initial roughness is 8.5648 µm. Assume 5 and 22 min of pre-cooling and post-cooling durations, respectively. (Answer: 82%)

Q2. What would be an increase in hardness of the ABS replica of the hip implant if the smoothing time is increased by 50% whereas the pre-cooling time is doubled? Assume required values. Take the surface hardness model as:

$$H_d = \left[\left(-0.004.T_{PT}^2 + 0.505.T_{PT} + 21.68\right).T_{PC}.\rho.R_{ai}^2\right]/T_s^4.$$

(Answer: 153%)

Q3. Calculate the percentage error in stem thickness of the hip replica as compared to CAD dimensions (6.6691 mm) when exposed to hot vapours for 10 s with pre-cooling and post-cooling durations of 25 and 10 min, respectively. The replica is manufactured at a 90° orientation angle with high density. The FDM produced a dimensional error of 1.4% when measured after fabrication. (Answer: 0.99%)

Q4. Calculate the hardness on the shore D scale of FDM parts having low density and exposed to hot chemical vapours for 25 s. There was a 92.45% improvement in surface finish with a 0.8432 µm average surface roughness after vapour smoothing. The parts are pre-cooled and post-cooled for 10 and 15 min, respectively. (Answer: 78.26)

References

1. Kumar, S. and Kruth, J.P. 2010. Composites by rapid prototyping technology. *Materials & Design*, 31:850–856.

2. Agarwala, M.K., Weeren, R.V., Bandyopadhyay, A., Whalen, P.J., Safari, A. and Danforth, S.C. 1996. Fused deposition of ceramics and metals: An overview. *Proceedings of the Solid Freeform Fabrication Symposium*, Austin, TX, pp. 385–392.

3. Onagoruwa, S., Bose, S. and Bandyopadhyay, A. 2001. Fused deposition of ceramics (FDC) and composites. *Proceedings of the Solid Freeform Fabrication Symposium*, Austin, TX, pp. 224–232.

4. Saengkwamsawang, P., Pimanpaeng, S., Amornkitbamrung, V., Rugmai, S. and Maensiri, S. 2011. Fabrication and characterization of Al-Al$_2$O$_3$ nanoparticles reinforced polyamide 6 composite filaments. *Proceedings of 18th International Conference on Composite Materials* 21st to 26th of August, 2011, in Jeju Island, Korea.

5. Masood, S.H., Mau, K. and Song, W.Q. 2010. Tensile properties of processed FDM polycarbonate material. *Materials Science Forum*, 654:2556–2559.

6. Negi, S., Dhiman, S. and Kumar Sharma, R. 2014. Basics and applications of rapid prototyping medical models. *Rapid Prototyping Journal*, 20:256–267.

7. Yang, F., Zhang, M. and Bhandari, B. 2017. Recent development in 3D food printing. *Critical Reviews in Food Science and Nutrition*, 57:3145–3153.

8. Choi, J., Kwon, O.C., Jo, W., Lee, H.J. and Moon, M.W. 2015. 4D printing technology: A review. *3D Printing and Additive Manufacturing*, 2:159–167.

9. Chhabra, M. and Singh, R. 2011. Rapid casting solutions: A review. *Rapid Prototyping Journal*, 17:328–350.

10. Omar, M.F.M., Sharif, S., Ibrahim, M., Hehsan, H., Busari, M.N.M. and Hafsa, M.N. 2012. Evaluation of direct rapid prototyping pattern for investment casting. *Advanced Materials Research*, 463:226–233.

11. Cooper, A.G., Kang, S., Kietzman, J.W., Prinz, F.B., Lombardi, J.L. and Weiss, L.E. 1999. Automated fabrication of complex molded parts using mold shape deposition manufacturing. *Materials & Design*, 20:83–89.

12. Singh, R., Singh, S. and Mahajan, V. 2014. Investigations for dimensional accuracy of investment casting process after cycle time reduction by advancements in shell moulding. *Procedia Materials Science*, 6:859–865.

13. Vasudevarao, B., Natarajan, D.P., Henderson, M. and Razdan, A. 2000. Sensitivity of RP surface finish to process parameter variation. *Proceedings of the Solid Freeform Fabrication Symposium*, Austin, TX, pp. 251–258.

14. Agarwala, M.K., Jamalabad, V.R., Langrana, N.A., Safari, A., Whalen, P.J. and Danforth, S.C. 1996. Structural quality of parts processed by fused deposition. *Rapid Prototyping Journal*, 2:4–19.

15. Chohan, J.S. and Singh, R. 2017. Pre and post processing techniques to improve surface characteristics of FDM parts: A state of art review and future applications. *Rapid Prototyping Journal*, 23:495–513.

16. Kattethota, G. and Henderson, M. 1998. A visual tool to improve layered manufacturing part quality. *Proceedings of Solid Freeform Fabrication Symposium*, Austin, TX, pp. 327–334.

17. Peng, A.H. 2012. Methods of improving part accuracy during rapid prototyping. *Advanced Materials Research*, 430:760–763.

18. Kumar, S., Kannan, V.N. and Sankaranarayanan, G. 2014. Parameter optimization of ABS-M30i Parts produced by fused deposition modeling for minimum surface roughness. *International Journal of Current Engineering and Technology*, 3:93–97.

19. Mani, K., Kulkarni, P. and Dutta, D. 1999. Region-based adaptive slicing. *Computer-Aided Design*, 31:317–333.

20. Pandey, P.M., Reddy, N.V. and Dhande, S.G. 2003. Slicing procedures in layered manufacturing: A review. *Rapid Prototyping Journal*, 9:274–288.

21. Bordoni, M. and Boschetto, A. 2012. Thickening of surfaces for direct additive manufacturing fabrication. *Rapid Prototyping Journal*, 18:308–318.

22. Leong, K.F., Chua, C.K., Chua, G.S. and Tan, C.H. 1998. Abrasive jet deburring of jewellery models built by stereolithography apparatus (SLA). *Journal of Materials Processing Technology*, 83:36–47.

23. Boschetto, A., Giordano, V. and Veniali, F. 2013. 3D roughness profile model in fused deposition modelling. *Rapid Prototyping Journal*, 19:240–252.

24. Pandey, P.M., Venkata Reddy, N. and Dhande, S.G. 2006. Virtual hybrid-FDM system to enhance surface finish. *Virtual and Physical Prototyping*, 1:101–116.

25. Chohan, J.S. and Singh, R. 2016. Enhancing dimensional accuracy of FDM based biomedical implant replicas by statistically controlled vapor smoothing process. *Progress in Additive Manufacturing*, 1:105–113.
26. Chohan, J.S., Singh, R. and Boparai, K.S. 2016. Parametric optimization of fused deposition modeling and vapour smoothing processes for surface finishing of biomedical implant replicas. *Measurement*, 94:602–613.
27. Garg, A., Bhattacharya, A. and Batish, A. 2016. On surface finish and dimensional accuracy of FDM parts after cold vapor treatment. *Materials and Manufacturing Processes*, 31:522–529.
28. Kuo, C.C. and Mao, R.C. 2016. Development of a precision surface polishing system for parts fabricated by fused deposition modeling. *Materials and Manufacturing Processes*, 31:1113–1118.
29. Chohan, J.S., Singh, R. and Boparai, K.S. 2016. Mathematical modelling of surface roughness for vapour processing of ABS parts fabricated with fused deposition modelling. *Journal of Manufacturing Processes*, 24:161–169.
30. Dym, C.L. 2004. *Principles of Mathematical Modeling*. New York: Elsevier.
31. Chohan, J.S., Singh, R., Boparai, K.S., Penna, R. and Fraternali, F. 2017. Dimensional accuracy analysis of coupled fused deposition modeling and vapour smoothing operations for biomedical applications. *Composites Part B: Engineering*, 117:138–149.

chapter five

Metal matrix composite from thermoplastic waste

Narinder Singh, Rupinder Singh and I. P. S. Ahuja

Contents

5.1   Introduction

The rapid growth of plastic/polymer consumption in different applications has inspired the research of innovative recycling procedures and improvement of former methods [1]. Polymer solid waste (PSW) is presenting new challenges and opportunities to the universe regardless of their technological advances and sustainability awareness [2]. The most common waste plastic accumulation is in municipal solid waste (MSW). Plastics in MSW are composed of various kinds of polymer waste such as HDPE, polyethylene terephthalate (PET), plastic films made of LDPE and hard plastic made of HDPE [3]. After single use, most of the plastic/polymer is discarded, resulting in large amounts of waste accumulation [4]. The percentage of plastic waste in landfills is going on the higher side that covers a huge space and takes hundreds of years for degradation [5]. That is why recycling of PSW becomes a need of the present day to sustain. Polymer recycling is a method for reduction of environmental problems caused by polymeric waste generation from day-to-day applications of polymer materials such as packaging and construction [6]. Various recycling methods are available in the present-day scenario; some of them are primary, secondary, tertiary and quaternary. Further, recycling methods can be subdivided into various techniques such as mechanical, chemical, incineration, etc. [7]. Amongst these techniques and methods, mechanical methods are more popular and simple—including extrusion, palletizing, drawing, etc. [8]. In the present work, a single screw extruder has been used for recycling of HDPE/LDPE collected from a local waste recycling plant and recycled to increase its application domain by preparing a filament wire through reinforcements of various ceramic particles in powder form (e.g. SiC and Al_2O_3). Figure 5.1 shows the schematic of a single screw extruder used for post-processing of the material.

Further, FDM has been used for preparing hybrid polymer composite for mechanical testing. The word 'hybrid composite' describes the composite containing more than one type of filler material as reinforcing additives or fillers [9]. One of the most common rapid prototyping (RP)/ additive manufacturing (AM) processes is FDM [10]. Rapid prototypes can be of invaluable help in testing the esthetic, functional and engineering performance of a product during its development cycle. In principle, any physical model should reproduce all relevant product properties at an accuracy level consistent with the intended evaluation stage [11].

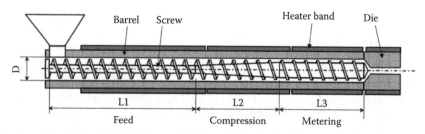

Figure 5.1 Schematic of single screw extruder.

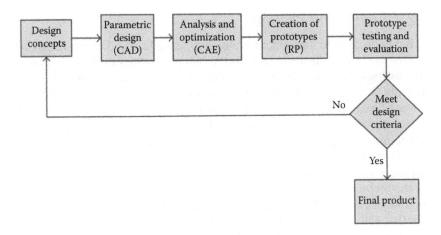

Figure 5.2 Product development cycle.

Rapid manufacturing is one of the manufacturing technologies used for fabrication of three-dimensional (3D) models using the layered manufacturing (LM) process by stacking and bonding thin layers [12]. Figure 5.2 shows the product development cycle for FDM.

LM processes are often used in various fields such as medical sciences, jewellery, construction, automobile, aircraft and others [13–15]. Creative production techniques have the advantage of manufacturing products by AM without the need for a forming tool [16]. It is also important to highlight that the wrong choice of process parameters may result in a lack of adherence between layers and even to the platform, resulting in warping of the object and faulty formation of layers [17]. As a consequence of the process characteristics, negative surfaces are generally produced by the deposition of a support material that is removed at the end of the manufacturing stage. This stage is called the post-processing stage [18–19]. It is important to note that even though in some cases the support material is water-soluble, the cost of the material is extremely high, making the process more expensive [17]. Commercially available RP methods include the following: stereolithography, selective laser sintering, FDM, laminated object manufacturing, ballistic particle manufacturing and 3D printing [20]. Among these, FDM is the most preferred and simple method for RP. In FDM, a feed stock filament wire is used, and usually that wire comprises acrylonitrile butadiene styrene (ABS). ABS has some specific properties, such as tensile strength, Young's modulus and percentage elongation. Many researchers have worked on parametric optimization of FDM with ABS as a filament [21]. In this study, a 3D printing machine form open-source vendor has been purchased and used for preparing the 3D printed parts.

5.2   Investment casting

IC has been used to manufacture weapons, jewellery and art castings during the ancient civilisation [22]. Today, its applications include jewellery/art castings, turbine blades and many more industrial/scientific components. It is known for its ability to produce components of excellent surface finish, dimensional accuracy and complex shapes. It is especially useful for making castings of complex and near-net shape geometry, where machining may not be possible or too wasteful. It is also considered to be the most ancient process of making art castings. Technological advances have also made it to be the most modern and versatile among all the metal casting processes [23]. Ancient man, initially, made the wax models of rudimentary tools. Few of such wax models were assembled together [24]. After the wax assembly was rammed inside a sand mould, it was heated up, which resulted in the wax draining out of the sand mould, leaving a hollow cavity inside. Molten metal was then poured into the cavity. The solidified weapons heads were then separated and finished. The survey clearly projects the basic principle, simplicity and efficiency of this IC process. Figure 5.3 shows the steps for the IC process.

The pattern in IC has the exact geometry of the required final cast part but with dimensional allowances to compensate its own volumetric shrinkage as well as the solidification shrinkage of the cast metal in the ceramic mould [25]. Steps in ICs have been discussed in Figure 5.3. In the first step an exact replica of the part has been prepared with the help of

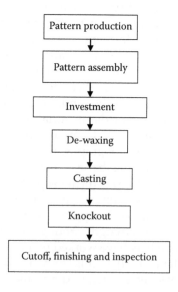

Figure 5.3 Steps in IC.

the 3D printer or any other method by using the material, which is able to melt and be removed. After the pattern preparation, a design tree is prepared and the whole tree is exposed to the slurry material, which is supposed to adhere to the surface of the pattern and prepare an outer protection layer. Slurry material is a kind of refectory material that is tenable to heat and does not melt but transfers the heat. After dipping the whole tree in slurry, it is given some time to dry. After the material has been dried off, the tree, along with the slurry layers, is exposed to the heat and material inside the slurry formation in the form. The final part is allowed to melt and come out, leaving an exact replica of the part. Finally, the molten metal is poured into the cavity and the metal is given some time to shape itself. After the material has cooled inside the cavity, the outer layers of the slurry layers are broken to get the final part. Wax, plastic, polystyrene or frozen mercury are the common pattern materials among which wax is most widely used [26]. Craig et al. [27] made extensive investigations on different waxes and concluded that a pattern wax must have the following characteristics:

1. It should have the lowest thermal expansion so that it can form a shape with the highest dimensional accuracy.
2. Its melting point should not be much higher than the ambient temperature so that distortion of thick sections and surface cavitations can be prevented.
3. It should be resistant to breakage – that is, it is of sufficient strength and hard enough at room temperature so that it can be handled without damage.
4. It should have a smooth and wet table surface so that a finished cast part with a smooth surface can be obtained.
5. It should have a low viscosity when melted to fill the thinnest sections of the die.
6. It should be released from the die easily after formation.
7. It should have a very low ash content so that it does not leave any ash inside the ceramic shell.
8. It should be environmentally safe – that is, it does not lead to the formation of environmentally hazardous or carcinogenic materials upon combustion.

Costs, availability, ease of recycling, toxicity, etc., are the other important factors when selecting a pattern wax. The working efficiency of IC can be increased by improving one or more characteristics of the wax, mentioned earlier, by mixing additives, blending with different waxes and varying the process parameters. The aforementioned information furnished by the authors can be highly useful and help the industries and researchers to select the right wax material for their work.

5.3   Metal matrix composites (MMC)

5.3.1   Preparation by various routes

An MMC is composite material with at least two constituent parts, one being a metal necessarily; the other material may be a different metal, such as a ceramic or organic compound [28]. When at least three materials are present, it is called a hybrid composite. An MMC is complementary to a cermet (any class of heat-resistant material). MMCs are made by dispersing a reinforcing material into a metal matrix [29]. The reinforcement surface can be coated to prevent a chemical reaction with the matrix. For example, carbon fibres are commonly used in aluminium matrix to synthesize composites showing low density and high strength. However, carbon reacts with aluminium to generate a brittle and water-soluble compound Al_4C_3 on the surface of the fibre [30]. To prevent this reaction, the carbon fibres are coated with nickel or titanium boride. The matrix is the monolithic material into which the reinforcement is embedded and is completely continuous. This means that there is a path through the matrix to any point in the material, unlike two materials sandwiched together [31]. In structural applications, the matrix is usually a lighter metal such as aluminium, magnesium or titanium, and it provides a compliant support for the reinforcement. In high-temperature applications, cobalt and cobalt–nickel alloy matrices are common. The reinforcement material is embedded into a matrix. The reinforcement does not always serve a purely structural task (reinforcing the compound), but it is also used to change physical properties such as wear resistance, friction coefficient or thermal conductivity. The reinforcement can be either continuous or discontinuous. Discontinuous MMCs can be isotropic, and they can be worked with standard metalworking techniques, such as extrusion, forging or rolling [32]. In addition, they may be machined using conventional techniques but commonly would need the use of polycrystalline diamond tooling (PCD). Continuous reinforcement uses monofilament wires or fibres such as carbon fibre or silicon carbide. Because the fibres are embedded into the matrix in a certain direction, the result is an anisotropic structure in which the alignment of the material affects its strength. One of the first MMCs used boron filament as reinforcement [33]. Discontinuous reinforcement uses 'whiskers', short fibres or particles. The most common reinforcing materials in this category are alumina and silicon carbide.

5.3.2   Methods to prepare MMC

The mechanical properties of MMCs are very sensitive to the method of processing being used. Considerable improvements may be achieved by applying science-based modelling techniques to optimize the processing procedure.

Following are some of the methods used to prepare MMC (Figure 5.4).

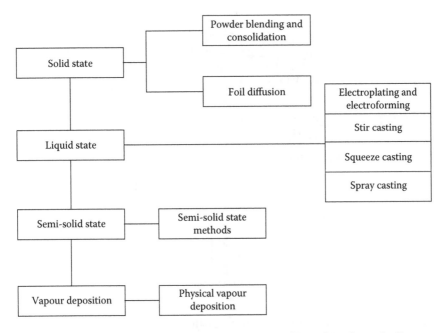

Figure 5.4 Various methods for MMC preparation. (From Liu, Q. et al., *Comput. Ind.*, 48, 181–197, 2002 [34].)

From aforementioned chart various methods for preparation of MMC can be seen. Liquid state methods are the widely used methods for the preparation of MMC. Here, two amongst them are discussed.

- *Stir casting*: Discontinuous reinforcement is stirred into molten metal, which is allowed to solidify.
- *Pressure infiltration*: Molten metal is infiltrated into the reinforcement through use of a kind of pressure such as gas pressure.
- *Squeeze casting*: Molten metal is injected into a form with fibres pre-placed inside it.
- *Spray deposition*: Molten metal is sprayed onto a continuous fibre substrate.

5.3.2.1 Stir casting

The vortex method is one of the better-known approaches used to create and maintain a good distribution of the reinforcement material in the matrix alloy [35]. In this method, after the matrix material is melted, it is stirred vigorously to form a vortex at the surface of the melt, and the reinforcement material is then introduced at the side of the vortex. The stirring

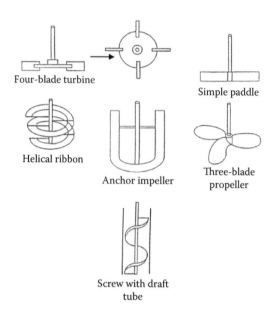

Four-blade turbine

Simple paddle

Helical ribbon

Anchor impeller

Three-blade propeller

Screw with draft tube

Figure 5.5 Various impellers used in stir casting.

is continued for a few minutes before the slurry is cast. Harnby et al. [36] studied different designs of mechanical stirrers, as shown in Figure 5.5. Among them, the turbine stirrer is quite popular. During stir casting for the synthesis of composites, stirring helps in two ways: (1) transferring particles into the liquid metal, and (2) maintaining the particles in a state of suspension. Several type of blades used in stir casting are shown in Figure 5.5.

The development of the vortex during stirring is observed to be helpful for transferring the particles into the matrix melt as the pressure difference between the inner and the outer surface of the melt sucks the particles into the liquid [37]. However, air bubbles and all the other impurities on the surface of the melt are also sucked into the liquid by the same mechanism, resulting in high porosity and inclusions in the cast product. A vigorously stirred melt will entrap gas, which proves to be extremely difficult to remove as the viscosity.

Gas injection of particles introduces a quantity of gas into the melt. Some of the methods, such as the ultrasonic, are very expensive and difficult to scale to production level [38]. Zero gravity processing is a very complicated method, and it is difficult to characterize. On account of centrifugal action, the distribution of the particles varies from the inner to the outer part of the billet because of the difference in centrifugal force [39]. Introducing the reinforcement particles from air to the stirred molten matrix will sometimes entrap the particles with other impurities, such as metal oxides and slag, which are formed on the surface of the melt. While pouring is done, air envelopes are also formed between the particles, altering the interface properties between the particles and the melt, and retarding the wet ability between them. In the case where the particles added are not at the same temperature as the slurry, the temperature and, consequently, the viscosity of the slurry will change vary rapidly. The particle distribution in cast composites may become inhomogeneous even when a homogeneous state of suspension is maintained in the slurry. During the solidification of a liquid matrix alloy containing dispersed second phase particles, the particles in the melt can migrate towards or away from the freezing front, and a particle near the freezing front will either be rejected or engulfed. These two phenomena lead to the redistribution of the particles during solidification. This means that the solidification cell size, and hence the solidification rate influence the distribution of the reinforcement particles in the final ingot. Fine dendrite arm spacing (DAS) produces a more uniform distribution of the particles, whereas larger DAS leads to particle clustering [40]. Rapidly solidified structures, therefore, give a better distribution of the particles due to finer dendrite size as well as due to a limited settling of the particles resulting from the reduced time during which the composites are in a molten state. A successful casting process must be able to produce a composite in which the particles are uniformly dispersed throughout the matrix [41]. The thoroughness of the agitation is determined by many factors, such as the shape of the agitator, its speed and its placement relative to the melt surface and the wall of the crucible. It is suggested that both the matrix and the reinforcement materials be preheated at a certain temperature before being mixed to release all the moisture and trapped air between the particles. The stirrer must be designed such that it avoids the agitation of the melt surface, and the formation of vortex must be avoided or minimized. The stirring speed should not be too high, but it should be continuous for a few minutes before the material is poured into a mould through the bottom of the crucible. Bottom pouring is necessary in order to avoid impurities on the surface of the melt being cast into the mould.

5.3.2.2   Pressure infiltration

Infiltration is a liquid state method of composite materials fabrication in which a preformed dispersed phase (ceramic particles, fibres, woven) is soaked in a molten matrix metal, which fills the space between the dispersed phase inclusions [42]. The motive force of an infiltration process may be either capillary force of the dispersed phase (spontaneous infiltration) or an external pressure (gaseous, mechanical, electromagnetic, centrifugal or ultrasonic) applied to the liquid matrix phase (forced infiltration). Infiltration is one of the methods of preparation of tungsten-copper composites.

The principal steps of the technology are as follows:

- Tungsten powder preparation with average particle size of about 1–5 mkm.
- Optional step: Coat the powder with nickel. Total nickel content is about 0.04%.
- Mix the tungsten powder with a polymer binder.
- Compact the powder by a moulding method (metal injection moulding, die pressing, iso-static pressing). Compaction should provide the predetermined porosity level (apparent density) of the tungsten structure.
- Solvent debinding.
- Sintering the green compact at 2200°F–2400°F (1204°C–1315°C) in the hydrogen atmosphere for 2 h.
- Place the sintered part on a copper plate (powder) in the infiltration/sintering furnace.
- Infiltration of the sintered tungsten skeleton porous structure with copper at 2100°F–2300°F (110°C–1260°C) in either the hydrogen atmosphere or vacuum for 1 h.

5.3.2.2.1 Gas pressure infiltration   Gas pressure infiltration is a forced infiltration method of liquid phase fabrication of MMCs, using a pressurized gas for applying pressure on the molten metal and forcing it to penetrate into a preformed dispersed phase (Figure 5.6) [43].

The gas pressure infiltration method is used for manufacturing large composite parts. The method allows for the use of non-coated fibres due to short contact time of the fibres with the hot metal. In contrast to the methods using mechanical force, gas pressure infiltration results in low damage of the fibres.

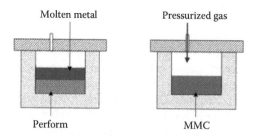

Figure 5.6 Steps involved in Pressure infiltration technique.

5.4 Preprocessing

In 3D printing, various types of wires such as ABS, PLA and others are being used nowadays [44]. These are easily available on the market. But in this study, 3D filament wire with specific application-based properties has been prepared with the use of a single screw extruder and by using HDPE as matrix material. Further, various reinforcements (such as SiC and Al_2O_3) have been made into matrix material, which is HDPE. For blending of the materials in different proportions, a twin screw extruder has been used.

Following are the steps involved in the process.

- Waste collection (HDPE in present case) and recycling.
- Cleaning of the waste polymer.
- Melt processing for pallet formation.
- Blending of the material in different proportions by using a twin screw extruder.
- Post-processing of material for final wire preparation.

Waste material is collected from local industry. After collection, material is sent for the recycling and pallets are formed. There are various methods of recycling of the polymers. Some of them are listed as follows.

Recycling methods (Figure 5.7)

- Primary recycling (mechanical recycling)
- Secondary recycling (mechanical recycling)
- Tertiary recycling (chemical processing)
- Quaternary recycling (incineration)

5.4.1 Primary recycling

Primary recycling, better known as re-extrusion, is the reintroduction of scrap, industrial or single-polymer plastic edges and parts to the extrusion cycle in order to produce products of similar material. This process utilizes scrap plastics that have similar features to the original products. Primary recycling is only feasible with semi-clean scrap, making it an unpopular choice with recyclers. Currently, most of the PSW being recycled is of process scrap from industry recycled via primary recycling techniques. In the UK, process scrap represents 250,000 tonnes of the plastic waste, and approximately 95% of it is primary recycled. Primary recycling can also involve the re-extrusion of post-consumer plastics. Generally, households are the main source of such waste stream. However, recycling household waste represents a number of challenges, namely the need of selective collection [45].

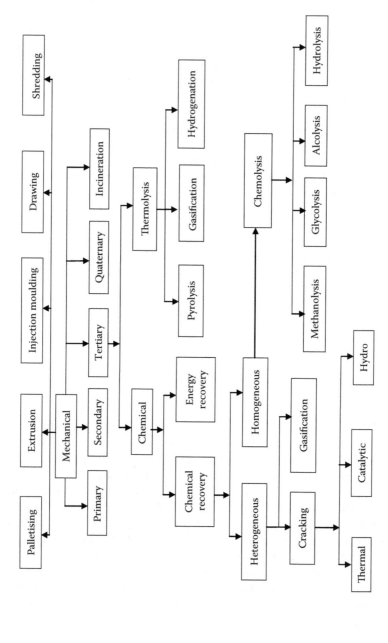

Figure 5.7 Different methods of recycling. (From Singh, N. et al., *Compos. B Eng.*, 115, 409–422, 2017 [45].)

5.4.2 Secondary recycling

Mechanical recycling is the process in which PSW is used in the manufacturing of plastic products via mechanical means, using recyclates, fillers and/or virgin polymers. Mechanical recycling of PSW can only be performed on single-polymer plastic. The more complex and contaminated the waste, the more difficult it is to recycle it mechanically. One of the main issues that face mechanical recyclers is the degradation and heterogeneity of PSW. Since chemical reactions that constitute polymer formation (i.e. addition, condensation, etc) are all reversible in theory, energy or heat supply can cause photo-oxidation and/or mechanical stresses, which occur as a consequence. Length or branching of polymer chains and/or cross-linking can also occur from the formation of oxidized compounds and/or harsh natural weathering conditions. A number of products in our daily lives come from mechanical recycling processes such as grocery bags, pipes, gutters, window and door profiles, etc. The quality is the main issue when dealing with mechanically recycled products. Extrusion profiling and quality of resultant products are the main concerns in utilizing thermoplastics in mechanical recycling in R&D projects [45].

5.4.3 Tertiary recycling

Several methods for chemical recycling are presently in use, such as direct chemical treatment involving gasification, smelting by blast furnace or coke oven, and degradation by liquefaction. Catalytic cracking and reforming facilitate the selective degradation of waste plastics. The use of solid catalysts such as silica-alumina, ZSM-5, zeolites and mesoporous materials for these purposes has been reported [46]. These materials effectively convert polyolefin into liquid fuel, giving lighter fractions as compared to thermal cracking. The main advantage of chemical recycling is the possibility of treating heterogeneous and contaminated polymers with limited use of pretreatment. If a recycler is considering a recycling scheme with 40% target or more, one should deal with materials that are very expensive to separate and treat. Thus, chemical recycling becomes a viable solution. Petrochemical plants are much greater in size (6–10 times) than plastic manufacturing plants. It is essential to utilize petrochemical plants in supplementing their usual feedstock by using PSW. Advanced thermochemical treatments of PSW in the presence of heat under controlled temperatures (thermolysis) provide a viable and an optimum engineering solution. Not only have they recovered healthy monomer fractions up to 60% in past reports, but they produce valuable petrochemicals that could be summarized as gases (rich with low-cut refinery products and hydrocarbons), tars (waxes and liquids very high in aromatic content) and char (carbon black and/or activated carbon).

Thermolysis processes can be divided into advanced thermochemical or pyrolysis (thermal cracking in an inert atmosphere), gasification (in the sub-stoichiometric presence of air usually leading to CO, CO_2 and H_2 production) and hydrogenation (hydrocracking). Appropriate design and scale (of operation and economy) are of paramount importance when it comes to thermal treatment plants. Thermal degradation behaviour in laboratory scale enables the assessment of a number of important parameters such as thermal kinetics, activation energy assessment (energy required to degrade materials treated and product formation) and determining reference temperatures of the half-life of polymers and maximum degradation point achievable. It is also important to perform pilot scale experiments utilising a number of rectors and unit operations before commencing with an alteration on a performance scale [47]. This will also aid in the determination of the mode of the material processing of the thermal plant (i.e. pulsating, continuous, batch, etc). Pyrolysis (depolymerization in inert atmospheres) is usually the first process in a thermal plant, and it is in need of appropriate end-product design. This could be achieved via the understanding of the systems kinetics on microscale (TGA) and pilot (bench scale unit) scale. Systems kinetics will not only develop appropriate models that will predict systems products and their interaction but through solving the derived mathematical expressions, it will predict the product interaction behaviour. This will assist in reducing side reaction and undesired by-products on an industrial scale. Developing rate expressions of the materials being treated will then be utilised in determining the optimum unit operation to be used and its required supply of power and proper media of operation – in the case of pyrolysis, the sufficient amount of inert atmosphere in the pyrolyser or the ratio of steam to oxygen in a gasifier.

5.4.4   Quaternary recycling

Economical constraints pose a major dilemma in industry, especially with recovery methods of processing scrap and heterogeneous waste streams. Energy recovery offers a solution to such problems by employing combustion processes to produce heat, steam and/or electricity. PSW possesses high calorific value when compared to other materials due to its crude oil origins. Since the heating value of plastics is high, they make a convenient energy source. Producing water and carbon dioxide upon combustion makes PSW similar to other petroleum-based fuels. In general, it is considered that incineration of PSW results in a volume reduction of 90%–99%, which reduces the reliability on landfilling [48]. In the process of energy recovery, the destruction of foams and granules resulting from PSW also destroys CFCs and other harmful blowing agents present.

5.5   A case study of an alternative method of developing MMC

Initially, the melt flow index test was performed (as per ASTM-D-1238 standard) at 190°C and 2.8 kg weight in order to select the appropriate proportions of alternative reinforced filament ingredients that should match the rheology of commercially used OEM material. Then, a single screw extruder (L/D ratio 20) was used to prepare filament wires of different proportions as given in Table 5.1. The filaments developed were then fed into the 3D printer to fabricate IC sacrificial patterns. However, any shape and size can be selected for the sacrificial pattern of IC. IC moulds were prepared by coating casting trees with refractory layers of silica. Autoclaving and baking were performed simultaneously at 800°C (keeping the pouring cup upright in order to lock the abrasive Al_2O_3 and SiC particles in the mould cavity). The molten Al-6063 matrix was then poured into the resulting cavity.

5.5.1   Melt flow index

Melt flow index is an analytical method used for determination of quality of polymer and flow properties. Various research studies showed the relationship between the MFI and different physical and chemical properties, such as viscosity, shear strength, molecular weight distribution and

Table 5.1 Pilot experimentation data

S. No.	(A) HDPE	(B) SiC (by weight % age)	(C) Al_2O_3 (by weight % age)	MFI (g/10 min)	Peak strength (KN/mm²)
1	0.700	0.150	0.150	23	12.56
2	0.700	0.000	0.300	22.5	12.66
3	0.800	0.100	0.100	24.3	11.5
4	0.750	0.125	0.125	23	12.03
5	0.900	0.050	0.050	26	10.3
6	0.750	0.200	0.050	22.8	12.5
7	0.825	0.050	0.125	25.3	11.05
8	0.700	0.300	0.000	22.6	13.6
9	0.850	0.000	0.150	25.4	11.3
10	0.750	0.050	0.200	23.5	12.06
11	0.850	0.150	0.000	25.4	10.98
12	0.825	0.125	0.050	25.6	10.89
13	1.0	0.000	0.000	26.8	13.01

shear rate. The MFI method is a measure of ability of a polymer or plastic into flow under certain conditions like pressure and fixed-size orifice. Dimensions of orifice, temperature and load pressure are specified by ASTM standards, and then value is measured in grams per 10 min, which is heated by the surrounding heater. The heater is covered by insulating material so that the heat loss can be minimized, and this heater is processed by automatic machine controls. This system consists of standard weights that pressurise the melted material from upper side and forces the material to flow out of die at standard condition. After selection of polymer waste as the matrix material the MFI of HDPE was tested with reinforcement of SiC and Al_2O_3 powder (Table 5.1).

In this experimentation work, an effort has been made to develop a feed stock filament wire that has different properties, which can be varied according to application. For this, a certain set of experiments were performed and analysed for different properties. Every experiment was performed with different proportions of HDPE-SiC-Al_2O_3. First of all, MFI was established for each composition/proportion of reinforcement. Finally, fixed proportions were taken for further processing on a single screw extruder (Table 5.1).

5.5.2 Single screw extruder

In the extrusion process, polymers are generally fed from a hopper that is gravimetric. The material goes through the hopper and comes in contact with the screw extruder. The screw is rotating at generally 10 rpm, forcing the plastic polymer in a straight direction inside the barrel. Screw rpm can be controlled by a controller unit that is provided with the machine. Material is heated by a heater mounted on a barrel, which can range from 120°C to 300°C depending on the polymer. Time of cooling and speed of rolling of the material play a major role in the properties of the wire being extruded. Three or more independent controlled heating zones gradually increase the temperature of the barrel from the rear end (where the plastic enters) to the front. This allows the plastic polymer to melt easily or effectively as it is forced through the screw rotation and reduces the risk of overheating, which may cause overheating of the polymer. When melted, material comes out of the die and can be rolled in form of wire easily by the rolling unit, which is available at the machine site as a unit of the machine. To maintain the uniformity of the extruded material, some arrangements are made to preheat the material [30]. The screw extrusion process has the following number of stages:

1. Solids conveying of material in the hopper.
2. Drag solids conveying in the initial turns of the screw.
3. Delay in melting, due to the development of a thin film of melted material separating the solids from the surrounding metallic wall(s).

4. Melting, where a specific melting mechanism develops, depending on the local pressure and temperature gradients.
5. Pumping involving the complex but regular helical flow pattern of the fluid elements towards the die.
6. Flow through the die.

Initially, a pilot study was conducted on a single screw extruder, and it was found that various factors such as barrel temperature, die temperature and screw speed contributed towards the properties of extrudate. In the pilot experimentation, design experiment software has been used to validate data (Table 5.1).

After getting the settings from the design expert, experimentation has been performed based on the setting and the values for the same have been obtained. After analysing the data, the ANOVA table has been obtained and shown in Table 5.2.

From Table 5.2, it can be clearly seen that P-values for all the parameter settings are less than 0.05 and can be considered significant. Hence the data provided in Table 5.2 is validated by the ANOVA by using the design expert software. Further, the plot has been obtained for MFI for its accuracy and found in line with the actual values (Figure 5.8).

Based on the data shown in Table 5.1 the following model has been developed and shown as follows:

$$MFI = +27.52 \times A + 22.67 \times B + 22.78 \times C \qquad (5.1)$$

Equation 5.1 has been drawn after analysis of the data obtained after the experimentation. After analysis a graph has also been obtained that showed the deviation of the predicted values form the actual value. The aforementioned procedure has been followed for the analysis of the peak

Table 5.2 ANOVA response

Source		Sum of squares	Df	Mean square	F-value	p-value Prob > F
Model		9.85	5	1.97	10.22	0.0041
Linear Mixture		2.15	2	1.07	5.57	0.0357
AB	5.07	1	5	0.07	26.31	0.0014
AC	2.22	1	2	0.22	11.50	0.0116
BC	0.075	1	0	0.075	0.39	0.5522
Residual		1.35	7	0.19		
Cor Total		11.20	12			

Note: Df denotes degree of freedom.

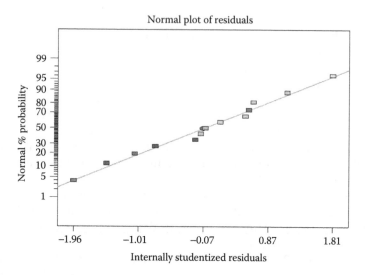

Figure 5.8 Plot for MFI.

strength. After analysis, the chart obtained shows the congruency of the predicted values with the actual value (Figure 5.9).

After analysing the values of peak strength from Table 5.1, a mathematical model has been obtained and shown in Equation 5.2.

$$\text{Peak strength} = +12.71 \times A + 13.74 \times B + 12.73 \times C - 9.32 \\ \times A \times B - 6.16 \times A \times C - 1.13 \times B \times C \tag{5.2}$$

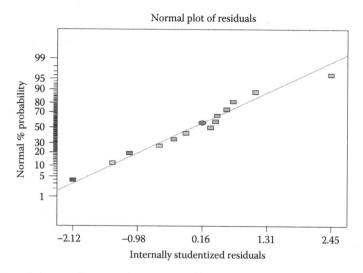

Figure 5.9 Plot for Peak strength.

5.5.3   Investment casting

Filament wires that were developed by using a single screw extruder were fired on an FDM system, and cubical patterns of size 20 × 20 × 20 mm were fabricated. It has been found that the surface finish of FDM-based IC patterns was poor and could affect the casting finish. IC ceramic moulds were prepared by coating the IC trees with refractory material. Once prepared, the ceramic moulds were autoclaved-cum-baked at 1150°C for 30 min (keeping the pouring cup upright in order to lock the abrasive-SiC and Al_2O_3 particles in the mould cavity). The molten Al-6063 matrix was then poured into the resulting cavity. First of all, as discussed earlier, 3D parts in the form of the cubes have been prepared in the 3D printer machine. After preparation of the cubical parts, the IC trees for the IC casting were prepared. After both parts and the IC trees were prepared, both were assembled to perform the IC process. After assembly of parts, the trees were firstly coated with ceramics. Casting tree assembly parts such as runner, pouring cup and gates were fabricated on the FDM system using HDPE material. Finally, the parts were assembled together, and the resulting casting tree is shown in Figure 5.10.

The casting trees were dipped into clay. The clay ingredients were mixed in a drum for about 72 h, which further mixed for 12 h to achieve the required clay viscosity of 60 s. After having a coat of clay, the IC trees were further coated with silica sand. These two steps (i.e. clay coating and silica coating) were repeated again and again, with an average interval of 4–5 h during which the coating was dried in an air-controlled environment (at 25°C and 65% humidity), until the desired mould sand thickness was attained. During stucco coating, the primary coat was made with zircon sand (grit size of 0.15 mm) so that the maximum surface finish on the counter casting surface could be achieved. The secondary, tertiary and quarter coating were made with 0.3, 0.5 and 0.65 mm grit size, respectively. Normally for Al casting, 5–7 layers are used depending upon the size and shape of geometry. But in the present research work, cracks in the ceramic mould (with 6 layers) were seen during the pilot experimentation, which might have been due to thermal expansion of plastic-based pattern material. After conducting a couple of experiments, the successful autoclaving of reinforced patterns was carried out on seven of the layers.

Figure 5.10 IC casting procedure.

Ash content

Ash content

Figure 5.11 Finally prepared castings.

Further, an increase in the number of layers has been investigated to study the effect of mould wall thickness on the mechanical properties of the Al-SiC-Al$_2$O$_3$-based composite castings. Prior to pouring, autoclaving of the reinforced pattern was carried out by heating the moulds in an electric furnace at 1150°C for 30 min. Moulds were kept in an upright position in order to arrest the abrasive SiC-Al$_2$O$_3$ particles inside the mould cavity, which otherwise may flow out. Figure 5.11 shows autoclaved ceramic moulds and final prepared castings.

It has been observed that small ash contents were presented in the mould cavity. Al-6063 alloy is one of the predominant metals in today's industrial practices and is used as a matrix metal in present research work.

5.6　Summary

In this chapter a novel method for the development of MMC with the help of HDPE and various ceramic reinforcements such as SiC and Al$_2$O$_3$, has been discussed in detail. Along with this, various other methods including stir casting, squeeze casting and pressure infiltration have been discussed. But in this case study, 3D printer patterns have been prepared as rapid patterns to be used for sacrificial IC. After successful preparation of the 3D printed parts in the form of the cube, parts were assembled with an IC tree and slurry layer were applied. Finally, those trees were exposed to heat and the remaining cavities were filled with the aluminium to get final MMC in the form of cube. It has been observed that reinforcements that were blended into the filament wire were present in the final casting, which is supposed to enhance the surface properties of the material. The rheological and mechanical testing of the filament wire has been done and analysed with the help of the design expert software, by getting an empirical model that validates the process.

By using those equations, a tailor-made wire with application-dependent filament wire can be prepared and further could be used in the larger domain.

Theoretical questions

Q1. Name the different methods to develop metal matrix composites.

Q2. What are the advantages of using additive manufacturing technology in IC?

Q3. How could reinforcements be blended with the matrix material?

Q5. What does the P-value signify in ANOVA?

Q6. Draw various diagrams of blades used in the case of stir casting.

Q7. Define the use of twin screw extruder and single screw extruder.

Q8. What are the advantages of using ceramic reinforcements in MMC?

Q9. List any five advantages of MMC over conventional material.

Q10. What are the controllable parameters in an FDM machine?

Q11. What are the units of MFI?

Q12. Explain the importance of MFI.

Q13. What is the importance of the layer thickness in the case of IC?

Q14. What is the melting temperature of aluminium?

Numerical questions (solved)

Q1. Find the value of the MFI if the matrix material is taken 50% and the rest of the material is SiC and Al_2O_3.
(Refer Table 5.1).
MFI follows following equation:

$$MFI = +27.52 \times A + 22.67 \times B + 22.78 \times C$$

Q2. A filament wire has to be prepared with the help of the FDM machine. What would be the peak strength of the wire if the matrix material is 60%, SiC is 20%, and the rest of the material is Al_2O_3?
Peak strength equation:

$$\text{Peak strength} = +12.71 \times A + 13.74 \times B + 12.73 \times C - 9.32$$
$$\times A \times B - 6.16 \times A \times C - 1.13 \times B \times C$$

References

1. Serranti, S., Luciani, V., Bonifazi, G., Hu, B., and Rem, P. C. (2015). An innovative recycling process to obtain pure polyethylene and polypropylene from household waste. *Waste Management*, 35, 12–20.
2. Al-Salem, S. M., Lettieri, P., and Baeyens, J. (2009). Recycling and recovery routes of plastic solid waste (PSW): A review. *Waste Management*, 29(10), 2625–2643.
3. Rigamonti, L., Grosso, M., Møller, J., Martinez Sanchez, V., Magnani, S., and Christensen, T. H. (2014). Environmental evaluation of plastic waste management scenarios. *Resources, Conservation and Recycling*, 85, 42–53.

4. Jayaraman, K., and Halliwell, R. (2009). Harakeke (phormium tenax) fibre-waste plastics blend composites processed by screwless extrusion. *Composites Part B: Engineering*, 40(7), 645–649.

5. Anuar Sharuddin, S. D., Abnisa, F., Wan Daud, W. M. A., and Aroua, M. K. (2016). A review on pyrolysis of plastic wastes. *Energy Conversion and Management*, 115, 308–326.

6. Hamad, K., Kaseem, M., and Deri, F. (2013). Recycling of waste from polymer materials: An overview of the recent works. *Polymer Degradation and Stability*, 98(12), 2801–2812.

7. Kumar, S., Panda, A. K., and Singh, R. K. (2011). A review on tertiary recycling of high-density polyethylene to fuel. *Resources, Conservation and Recycling*, 55(11), 893–910.

8. Campbell, G., and Spalding, M. (2013). Single-screw extrusion: Introduction and troubleshooting. *Analyzing and Troubleshooting: Single Screw Extrusion*, pp. 1–22. Cincinnati, OH: Carl Hanser Publishers.

9. Kumar, R., Singh, T., and Singh, H. (2015). Solid waste-based hybrid natural fiber polymeric composites. *Journal of Reinforced Plastics and Composites*, 34(23), 1979–1985.

10. Espalin, D., Alberto Ramirez, J., Medina, F. et al. (2014). A review of melt extrusion additive manufacturing processes: I. Process design and modeling. *Rapid Prototyping Journal*, 20(3), 192–204.

11. Armillotta, A. (2006). Assessment of surface quality on textured FDM prototypes. *Rapid Prototyping Journal*, 12(1), 35–41.

12. Ahn, D., Kweon., J. H., Kwon, S., Song, J., and Lee, S. (2009). Representation of surface roughness in fused deposition modeling. *Journal of Materials Processing Technology*, 209(15–16), 5593–5600.

13. Chua, C. K., Leong, K. F., and Lim, C. S. (2003). *Rapid Prototyping: Principles and Applications*. Singapore: World Scientific.

14. Gebhardt, A., and Gibson, I. (2003/2006). Rapid prototyping: From product development to medicine and beyond. *Virtual and Physical Prototyping*, 1, 31–42.

15. Venuvinod, P. K., and Ma, W. (2004). *Rapid Prototyping: Lased Based and Other Technologies*. Dordrecht, the Netherlands: Kluwer Academic Publishers.

16. Bagsik, A., and Schöoppner, V. (2011). Mechanical properties of fused deposition modeling parts manufactured with Ultem 9085. In: *Proceedings of the 69th Annual Technical Conference of the Society of Plastics Engineers (ANTEC'11)*, Boston, MA, pp. 1294–1298.

17. Cunico, M. W. M. (2013). Study and optimisation of FDM process parameters for support-material-free deposition of filaments and increased layer adherence. *Virtual and Physical Prototyping*, 8, 127–134.

18. Volpato, N. (Ed.). (2007). *Prototipagem rápida: Tecnologias e aplicações (Rapid prototyping Technologies and Applications)*. São Paulo, Brazil: Edgard Blucher.

19. Gibson, I., Rosen, D. W., and Stucker, B. (2010). *Additive Manufacturing Technologies: Rapid Prototyping to Direct Digital Manufacturing*. New York: Springer.

20. Pham, D. T., and Gault, R. S. (1998). A comparison of rapid prototyping technologies. *International Journal of Machine Tools and Manufacturing*, 38(10/11), 1257–1287.

21. Panda, S. K. (2009). Optimization of fused deposition modelling (FDM) process parameters using bacterial foraging technique. *Intelligent Information Management*, 1(2), 89–97.
22. Pattnaik, S., Karunakar, D. B., and Jha, P. K. (2012). Developments in investment casting process—A review. *Journal of Materials Processing Technology*, 212(11), 2332–2348.
23. Groover, M. P. (2007). *Fundamentals of Modern Manufacturing: Materials Processes, and Systems*. Hoboken, NJ: John Wiley & Sons.
24. Heard, C. (2009). Uneasy Associations: Wax Bodies Outside the Canon. In: Boldt-Irons, L. A., Federici, C., and Virgulti, E. (Eds.), *Disguise, Deception, Trompe-l'œil: Interdisciplinary Perspectives*, pp. 231–250. New York: Peter Lang.
25. Chattopadhyay, A. B. (2004). Finishing of critical components with high surface integrity. In: *Proceedings of the National Conference on Investment Casting: NCIC 2003*, p. 29. New Delhi, India: Allied Publishers.
26. Pattnaik, S., Karunakar, D. B., and Jha, P. K. (2014). A prediction model for the lost wax process through fuzzy-based artificial neural network. *Proceedings of the Institution of Mechanical Engineers, Part C: Journal of Mechanical Engineering Science*, 228(7), 1259–1271.
27. Craig, H., and Gordon, L. I. (1965). Deuterium and oxygen 18 variations in the ocean and the marine atmosphere. In: Tongiorgi, E. (Ed.), *Stable Isotopes in Oceanographic Studies and Paleotemperatures*, pp. 9–130. Pisa, Italy: Laboratorio di Geologia Nucleare.
28. Clyne, T. W., and Withers, P. J. (1995). *An Introduction to Metal Matrix Composites*. Cambridge, UK: Cambridge University Press.
29. Ibrahim, I. A., Mohamed, F. A., and Lavernia, E. J. (1991). Particulate reinforced metal matrix composites—A review. *Journal of Materials Science*, 26(5), 1137–1156.
30. Garg, R. K., Singh, K. K., Sachdeva, A., Sharma, V. S., Ojha, K., and Singh, S. (2010). Review of research work in sinking EDM and WEDM on metal matrix composite materials. *The International Journal of Advanced Manufacturing Technology*, 50(5–8), 611–624.
31. Datta, S. (1997). *Electronic Transport in Mesoscopic Systems*. Cambridge, UK: Cambridge University Press.
32. Ramesh, C. S., Adarsha, H., Harishanad, K. S., and Naveen Prakash, N. (2012). A review on hot extrusion of metal matrix composites (MMC's). *International Journal of Engineering and Science*, 1(10), 30–35.
33. Singh, H., Sarabjit, N. J., and Tyagi, A. K. (2011). An overview of metal matrix composite: Processing and SiC based mechanical properties. *Journal of Engineering Research and Studies*, 2, 72–78.
34. Liu, Q., Sui, G., and Leu, M. C. (2002). Experimental study on the ice pattern fabrication for the investment casting by rapid freeze prototyping. *Computers in Industry*, 48, 181–197.
35. Hashim, J., Looney, L., and Hashmi, M. S. J. (1999). Metal matrix composites: Production by the stir casting method. *Journal of Materials Processing Technology*, 92, 1–7.
36. Harnby, N., Edward, M. F., and Nienow, A. W. (1985). *Mixing in Process Industries*. London, UK: Butterworths.
37. Girot, F. A., Albingre, L., Quenisset, J. M., and Naslain, R. (1987). Rheocasting Al matrix composites. *Journal of Metals*, 39, 18–21.

38. Hashim, J., Looney, L., and Hashmi, M. S. J. (2002). Particle distribution in cast metal matrix composites—Part I. *Journal of Materials Processing Technology*, 123(2), 251–257.

39. Lajoye, L., and Suery, M. (1987). Centrifugal casting of aluminum alloy matrix composites. In: *Proceedings of the Conference on Solidification Processing*, Sheffield, UK, September, pp. 473–476.

40. Samuel, A. M., Gotmare, A., and Samuel, F. H. (1995). Effect of solidification rate and metal feedability on porosity and SiC/Al$_2$O$_3$ distribution in an Al–Si–Mg (359) alloy. *Composites Science and Technology*, 53, 301–305.

41. Ma, P. C., Siddiqui, N. A., Marom, G., and Kim, J. K. (2010). Dispersion and functionalization of carbon nanotubes for polymer-based nanocomposites: A review. *Composites Part A: Applied Science and Manufacturing*, 41(10), 1345–1367.

42. Chou, T. W., Kelly, A., and Okura, A. (1985). Fibre-reinforced metal-matrix composites. *Composites*, 16(3), 187–206.

43. Chawla, K. K. (2006). *Metal Matrix Composites*. Weinheim, Germany: Wiley-VCH Verlag GmbH & Co. KGaA.

44. Gross, B. C., Erkal, J. L., Lockwood, S. Y., Chen, C., and Spence, D. M. (2014). Evaluation of 3D printing and its potential impact on biotechnology and the chemical sciences. *Analytical Chemistry*, 86, 3240–3253.

45. Singh, N., Hui, D., Singh, R., Ahuja, I. P. S., Feo, L., and Fraternali, F. (2017). Recycling of plastic solid waste: A state of art review and future applications. *Composites Part B: Engineering*, 115, 409–422.

46. Corma, A. (1997). Preparation and catalytic properties of new mesoporous materials. *Topics in Catalysis*, 4(3–4), 249–260.

47. Al-Salem, S. M., Lettieri, P., and Baeyens, J. (2010). The valorization of plastic solid waste (PSW) by primary to quaternary routes: From re-use to energy and chemicals. *Progress in Energy and Combustion Science*, 36(1), 103–129.

48. Al-Salem, S. M., Lettieri, P., and Baeyens, J. (2009). Recycling and recovery routes of plastic solid waste (PSW): A review. *Waste Management*, 29(10), 2625–2643.

chapter six

Joining of dissimilar thermoplastic with friction stir welding through rapid tooling

Ranvijay Kumar, Rupinder Singh and I. P. S. Ahuja

Contents

6.1 Introduction

Thermoplastics are characterized by their nature of exhibiting physical, chemical, mechanical, thermal and morphological behaviour such as tensile strength, elongation, impact strength, porosity, hardness, amorphous versus crystalline, melting and glass transition temperature, surface behaviour, carbon chain length, molecular arrangement and molecular weight versus molecular density. Dissimilar thermoplastic material possesses non-compatibility issues that hinder it from being used in different applications, especially where application of thermoplastic needs to be compatible by characteristics (e.g. solid state welding) [1–5]. Some of the studies that have focussed on these issues have revealed the mechanism, theory and techniques by which two dissimilar thermoplastic materials can be characterized – for example, reinforcement of metallic and non-metallic fillers leads to the improvement in the mechanical and morphological characteristics of the polymer matrix as well as polymer compatibility for joining and mixing [6–8]. For the specific cases of the friction welding applications, FDM can be used for the fabrication of the consumable tool that will be used for joining polymeric sheets. For example, pipelines are generally made up of thermoplastic material, and the occurrence of cracks and leakages are the most common issues that cause material loss, time loss and economic loss. To prevent these problems, a rapid tool such as a drill bit can be applied to fix those issues [9–13]. For a $50 \times 30 \times 4$ mm sheet of HDPE, the FWS is eligible to weld at a tool rotational speed of 1500, 2100 and 3000 rpm; a tool transverse speed at 45, 75 and 115 mm/min; and a tilt angle of 1°, 2° and 3°. Tool rotational speed contributed the maximum for the change in the output as 73.85% [14]. Squeo et al. [15] have conducted similar operations. For a $150 \times 60 \times 6$ mm sheet of MDPE, the rotational speed between 1400 and 2000 rpm, tilt angle at 1°–2° and travel speed of 15 mm/min is useful. The rational speed contributes the maximum for the change in the elongation, and the tilt angle contributes the maximum for the change in the tensile strength of the MDPE joints produced by FSW [16]. A $100 \times 25 \times 3$ mm PMMA specimen requires 500–200 rpm, 5.5–12 s welding time, joining pressure of 3 bar and plunge depth of 3.5–4 mm [17,18]. Friction-based Injection Clinching Joining (F-ICJ) technique for production of a hybrid structure of PEI-AA6082-T6 requires ultra-high rotational speed in between 7500 and 20,000 rpm, frictional force of 0.2–0.5.Mpa and frictional time of 2500–5000 ms [19]. For the production of a 2 mm thick weld of carbon reinforced PA66 laminates by the friction spot welding, it is required that there be 1000–3000 rpm, 1.8–2.2 plunge depth and 3.6–7.5 s frictional time [20]. FSW for 10 mm PMMA plates requires a 700 rpm rotational speed, advancing speed of 25 mm/min, 15–35 mm shoulder diameter and 5–6 mm pin diameter [18]. For $100 \times 100 \times 3$ mm with a curve thickness of 2.7 mm dimension of AA6082-PP sheets, it is required that there be a rotational speed of 1000 rpm, feed rate of 100 mm/min, 2° tilt angle, 3000N force and 3–7 mm pin diameter [21]. For the joining of PC-PC sheets of $90 \times 20 \times 3$ mm dimension through FSW, there must be a tool plunge rate between 8–46 mm/min, tool rotation of 1500–5400 rpm, preheating time of 0–20 s, dwell time of 0–20 s and waiting time of 0–20 s for the joining to be feasible [22]. Welding joints are highly characterized by their micro-structural and mechanical properties, and sustainability of welded joints are investigated by thermal, mechanical, tribological and morphological properties [23–27]. Welding of Al to polyvinyl chloride possessed good shear strength sustainability of 16.1 Mpa, which was almost 75% to the parent PVC [28]. Mechanical sustainability of the joints is highly domi-nated by the role of tool geometry and configuration. As FSW conducted for polypropylene and polyethylene, the optimum pin geometry resulted in the tensile strength of 98% to the polyethylene as well as elongation and hardness greater than the base material [24]. High-density polyeth-ylene with dimensions of $100 \times 200 \times 10$ possessed 96% flexural strength as compared to base material at optimum processing conditions (RPM of 1400, transverse speed of 25 mm/min and shoulder temperature of 100° centigrade) [29].

The reported literature has been reviewed, and it was found that most of the studies have been reported regarding thermoplastic with use of the metallic and non-metallic fillers and their processing techniques. Some limited studies have reported that use of the reinforcement in the ther-moplastic matrix can be used to achieve the compatibility of the two dis-similar thermoplastics. There is a research gap found in order to fabricate the functional prototypes, which fulfils the compatibility of two polymers in terms of joining/welding. Compatibility of two dissimilar polymers enables the friction welding for maintenance and structural applications. The welded pieces, by introducing rapid tooling, have been investigated by measuring the mechanical properties such as peak load, peak strength, break load, break strength, break elongation, Shore D hardness, tool con-sumption rate and morphological property as percentage porosity at joints. Taguchi L9 orthogonal arrays as design of experiment supported by Minitab software have been used for the optimisation of an input pro-cess variable with applying a statistical model.

6.2   FSW

The FSW technique was developed by The Welding Institute in the UK in 1991 under the solid-state welding techniques. Friction stir welding and friction stir processing are similar processes but differentiated by the nature of application. FSW is a technique in which a non-consumable tool of rotating type is inserted in between the interfaces of two sheets to cre-ate the diffusion of their edges for joining of materials, whereas friction stir processing (FSP) is considered the advancement in the technique used for refinement of microstructural properties only. This FSP technique is not considerable for the joining purposes; friction stir processing is appli-cable to only where refinement of microstructures is required (Figure 6.1).

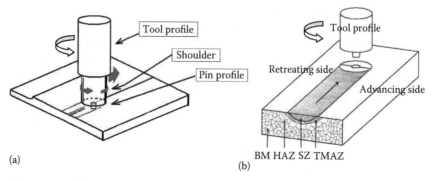

Figure 6.1 Schematic of (a) FSW, (b) FSP.

Advancing side in friction stir welding/processing can be considered as a side where flow of materials has higher temperature whereas the retreading side is the phase where material flow has a lower temperature. FSW and FSP can be differentiated by investigating advancing and retreading side [30–32]. The basic requirement of all solid-state welding processes is considered as heat generation much below the melting point of baseplates. The FSW process overcomes the problem caused during arc welding. FSW requires a shoulder that is protruded with a pin called a probe. The shoulder arrangement creates the frictional heating, and a pin probe facilitates the stirring to the joint interface (Figure 6.1) The friction welding tool (FSW tool) is one of the most important parts for success of the process. The tool consists of a rotary round shoulder and a threaded type of cylindrical pin. Although prominent efforts have been made in the previous years to develop economic and reusable tools, most of the attempts have been empirical in nature, and further work is needed for refinement in tool geometry to modify the practice of FSW to hard metals and alloys.

6.3   Case study for welding of dissimilar thermoplastic using rapid tooling

To introduce the influences of input process variables on welded pieces by rapid tooling, a case study has been conducted. The case study has been processed by use of different technologies, namely establishing MFI, differential scanning calorimetory (DSC), twin screw extrusion (TSE), fused deposition modelling (FDP) as 3D printing for rapid tooling preparation, FSW on milling setup and evaluation of mechanical and morphological properties.

6.3.1 Materials and methods

For the present study, ABS and PA6 polymer have been selected. ABS and PA6 are two different characterized polymers based upon the thermal, mechanical, chemical and metallurgical properties. Commercial-grade ABS (Grade EX58) and PA6 (Grade PX99848) were purchased from the Batra Polymers Pvt Ltd, India. The ABS and PA6 granules were tested through the DSC to check the melting characteristics. The mechanical properties were evaluated by a universal tensile tester (ASTM D628), a melt flow indexer (ASTM D1238) was used to evaluate the MFI, and a differential scanning calorimetric setup was used to evaluate the melting point. The mechanical, melt flow and thermal properties of virgin ABS and PA6 have been tabulated in Table 6.1.

Aluminium metal powder of 300 mesh size (approx. 50 μm grain size) was selected for the reinforcement media. The Al metal powder was induced in the ABS and PA6 polymer matrix to modify the MFI. The FSW process is aimed to be processed with a substrate of ABS composite and rapid tool of PA6 composite. Figure 6.2 shows the process of FSW for present case.

Table 6.1 Materials property of virgin ABS and PA6

Material	Peak load (Kgf)	Peak strength (MPa)	Shore D hardness	Melt flow index (g/10 min)	Melting point (°C)
ABS	59.2	23.40	73.0	8.76	201.22
PA6	203.0	94.90	71.5	23.27	219.35

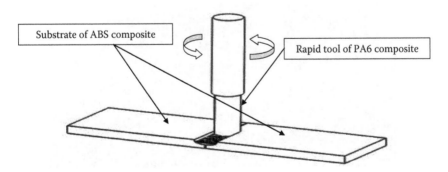

Figure 6.2 FSW process using rapid tool.

6.3.2   MFI characterization

The MFI represents the material flow behaviour and quality of thermoplastic materials. The ASTM D1238 is applicable for most of the thermoplastic materials. The 3.80 kg load is applied at 230°C through piston, and material is collected per 10 min for determination of melt flow index (Figure 6.3).

The melt flow properties of any thermoplastic material can be improved and modelled for different field application by adding certain proportions of fillers such as metallic and non-metallic powders, especially for friction welding application where any two dissimilar plastic-based material can join by maintaining a similar melt flow index. Table 6.2 shows that the MFI changes considerably by varying the proportion of Al metal powder (commercial, 50 micron grain size) in ABS and PA6 polymer matrix.

The MFI of ABS and PA6 with varying the Al from 5% to 50% were checked by following the ASTM D1238. The MFI of ABS at 15% Al content was found to be 11.56 g/10 min whereas the MFI of PA6 at 50% Al content was 11.96 g/10 min (Figure 6.4), which was under the very similar range as compared to the other proportions. So these proportions (namely ABS-15Al and PA6-50Al) were judicially selected for the feedstock filament preparations and FDM part fabrication for the friction welding feasibility.

It has been observed that in the case of the ABS polymer matrix by increasing the Al metal filler up to 20%, lead to increase in the MFI, but above this level there was a significant decrease in the MFI (Figure 6.3). It should be noted that molecular weight and density are the two different polymer properties that define the characteristics of polymers. The density is an inherent property of polymer matrix, which doesn't ascertain the

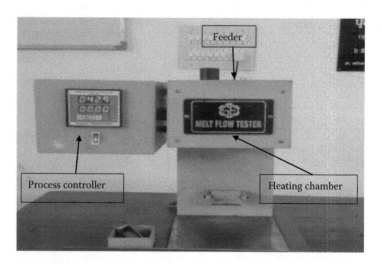

Figure 6.3 MFI tester.

Table 6.2 MFI variation of ABS and PA6 with
proportions of Al metal powder filler

Trial no.	Pure ABS	5%	10%	15%	20%	25%	30%	35%	40%	45%	50%
1	9.05	8.11	10.54	11.44	12.47	10.38	10.26	9.37	9.33	7.23	7.09
2	8.39	10.39	10.57	11.8	12.49	11.12	10.59	10.63	8.87	7.43	6.81
3	8.86	9.71	9.27	11.48	12.98	10.98	10.01	9.98	9.23	8.04	5.56
Avg.	8.76	9.40	10.12	11.57	12.64	10.82	10.28	9.99	9.14	7.56	6.48

Trial no.	Pure PA6	5%	10%	15%	20%	25%	30%	35%	40%	45%	50%
1	23.58	25.17	29.82	33.05	35.23	38.87	36.85	31.82	20.23	15.02	10.76
2	23.27	23.01	30.56	30.93	34.54	33.64	35.39	29.86	21.11	19.27	12.33
3	22.98	25.4	29.09	30.89	33.45	35.53	36.88	28.54	23.23	16.94	12.84
Avg.	23.27	24.52	29.82	31.62	34.40	36.01	36.37	30.07	21.52	17.07	11.97

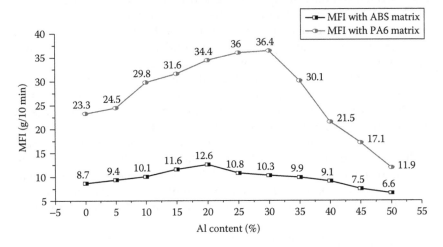

Figure 6.4 MFI characterizations with Al content.

determination of melt flow rate. Molecular weight is an interesting aspect
of polymer property, which practically ensures the melt flow rate. A higher
molecular weight ensures lower melt flow characteristics. As a polymer's
long carbon chain is branched, it causes the increase in molecular weight,
but linear or no branched carbon chain results in lower molecular weight.
In the present study, the MFI of the ABS matrix is increased up to the 20%
Al content (by weight), which may be due to the effect of density. But hence-
forth the melt flow was decreased mainly due to an increase in the molecular weight. This may be because of carbon chain branching/elongation
with the effect of Al powder reinforcement beyond 20% (by weight) under

the present environmental conditions. For the PA6 polymer matrix, the increase was observed in the MFI up to 30% of the Al filler but after that increase in the Al fillers, the MFI decreased.

6.3.3 DSC for thermal analysis

Differential scanning calorimetry (DSC) measures temperatures and heat flows associated with thermal transitions in a material. Common usage includes investigation, selection, comparison and end-use performance evaluation of materials in research, quality control and production applications. Properties measured by TA Instruments' DSC techniques include glass transitions, 'cold' crystallization, phase changes, melting, crystallization, product stability, cure/cure kinetics and oxidative stability. For thermal analysis, METTLER TOLEDO, Model DSC3, Swiss make with $STAR^e$ (SW 14.00) software was used in a N_2 gas environment. The typical DSC setup determines the behaviour of applied samples by taking references from a standard sample, both enclosed in a metallic crucible (Al or platinum). As shown in Figure 6.5, the DSC sensor uses two crucibles for heating and cooling, one for reference and another for sample.

The DSC evaluation was performed under controlled experimental environment conditions of continuous heating (endothermic changes, 10°C/min) and continuous cooling (exothermic changes, 10°C/min) in the 30–250°C temperature range through two consecutive cycles at 50 mL/min of a N_2 gas supply. The DSC was performed under the two cycles because after the first heating cycle the material becomes stable in nature and results in the correct material information. In the second cycle of heating, the peak melting of natural ABS and PA6 was obtained as 201.22°C and 219.35°C (Figure 6.6). For the peak melting of ABS and PA6, there was a large gap of melting point found, but the melting point of ABS-15Al and PA6-50Al was obtained considerably similar as 218.11°C and 218.27°C, respectively.

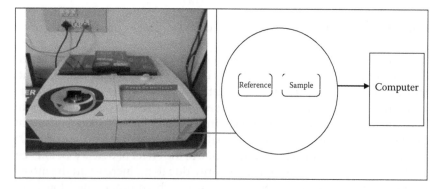

Figure 6.5 DSC setup and heating chamber for sample.

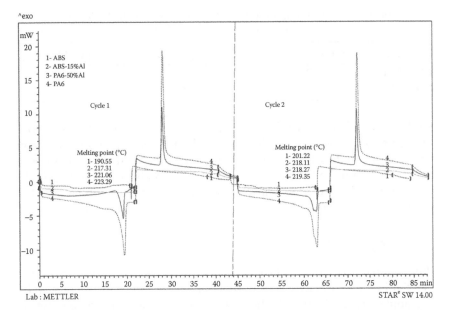

Figure 6.6 DSC plots for virgin ABS[1], ABS-15Al[2], PA6-50Al[3] and virgin PA6[4].

It justified the hypotheses that definite proportions of Al content leading to similar MFI also contributed to their similarities in the melting range.

6.3.4 Twin screw extrusion

Twin screw extrusion (TSE) is one of the most advanced techniques for the preparations of the feedstock filament and for uniform mixing of the reinforcement agent with a polymer matrix. The main advantages of twin screw extruders (intermeshing corotating) are their exceptional mixing capability that gives the remarkable characteristics to extruded products. In the present study, an intermeshing co-rotating type of TSE has been selected for experimentations. The TSE is operating between the temperature range of 25°C–400°C, rotational speed range of 0–300 rpm and torque range of 0–2 nm. The nozzle can be changed from 0.25 to 2.0 mm in diameter. As required of applications, the pilot experimentation was conducted and the input process parameter was decided based upon the uniformity of the filament extruded. Except for temperature, the nozzle diameter and operating torque were the two other machining parameters that have significant effect on the change in the properties of the obtained feedstock filaments. The temperature for extrusion has been selected as the DSC results obtained (218.11°C for ABS-15Al and 218.27°C for PA6-50Al). For better mixing of the Al with ABS and a PA6 matrix, the extrusion temperature should be greater than the melting point, so the temperature of 220°C for ABS-15Al and 245°C

Table 6.3 Processing setup for preparation of feedstock filaments of ABS-15Al and PA6-50Al

Materials	Temperature (°C)	Screw speed (rpm)	Applied load (kg)
ABS-15Al	220	30	20
PA6-50Al	245	20	15

for PA6-50Al) has been selected. The acceptable diameter for feedstock filaments to be compatible with FDM is 1.5 mm, so the nozzle diameter has been selected as 1.5 mm. The screw speed and applied load have been selected based upon the pilot experimentation conducted. Based upon the uniformity achieved, the following set of parametric combinations has been selected for ABS-15Al and PA6-50Al, as given in Table 6.3.

6.3.5 3D printing

Prepared feedstock filaments of ABS-15Al and PA6-50Al were fed to a commercial FDM set to fabricate non-functional prototypes for friction welding applications. Figure 6.7 shows the configuration of the commercial FDM setup and parts fabricated in the form of pin and rectangular sheets.

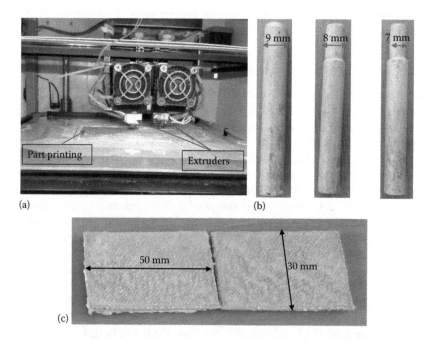

Figure 6.7 Photographic view of (a) part fabrication on FDM, (b) printed shoulder less pin of PA6-50Al for FSW and (c) printed sheets of ABS-15Al for joining.

FDM processed these parts under maintaining the same input process parameters. Under an infill density of 0.8, 6 number of perimeter, deposition angle of 60°, nozzle diameter of 0.3 mm, filament diameter of 1.75 mm, honeycomb fill pattern, perimeter speed of 30 mm/s, infill speed of 60 mm/s, travel speed of 130 mm/s, extruder temperature of 250°C and bed temperature of 55°C, parts were prepared. A rectangular sheet was prepared with ABS-15Al whereas PA6-50Al was used to fabricate the pin profile under these same processing conditions.

6.3.6 *Taguchi-based design of experimentation*

Pilot experimentation was performed under the 5600–2000 rpm, transverse speed of 10–100 mm/min and pin diameter of 2–20 mm. Under a rotational speed of 1000–1500 rpm, welding of sheets appeared as good strength but out of this range resulted in the poor welding appearance, so based upon this variability, FSW was processed. Similarly, the transverse speed range out of 30–50 mm/min resulted in the poor welding appearance, so 30–50 mm/min has been selected as the final for the present case. A pin diameter from 0 to 6 was not able to resist the thrust force at 1000–1500 rpm and 30–50 mm/min of the transverse speed and pin was broken during the welding process. Similarly, the pin diameter of 10–20 mm produced the larger welding zone, which promoted the poor joining strength. Considering these issues, a pin diameter of 7–9 mm has been selected for the present study. For preparation of design of experimentation, A design of experiment with three factor and three level has been selected for input process parameters. Considering these input variables, orthogonal arrays of Taguchi L9 have been developed as tabulated in Table 6.4.

Table 6.4 Design of experiment based upon Taguchi L9 orthogonal arrays for FSW process

Exp no.	Rotational speed (rpm)	Transverse speed (mm/min)	Pin diameter (mm)
1	1000	30	7
2	1000	40	8
3	1000	50	9
4	1200	30	8
5	1200	40	9
6	1200	50	7
7	1400	30	9
8	1400	40	7
9	1400	50	8

6.3.7 FSW

The commercial vertical milling machine with required changes in the configurations has been installed for FSW processing. The experimental setup configured with rotational speed of 0–2000 rpm and transverse speed of 5–150 mm/min. Both rotational speed and transverse speed were automatically controlled to achieve the maximum accuracies during the process. The vertical milling machine was modified in such a way that table movement can be directed in both orientations – that is, from right-left or left-right, and up-down or down-up. Figure 6.8 shows the experimental setup and the fixture design that was used for the FSW processes.

For the final welding experimentations, two sheets of ABS-15Al with dimensions of 50 × 30 × 4 mm were clamped in fixture and pin of PA6-50Al was acted upon the sheets to process FSW. The table movement was operated from left to right, and pin rotation was kept in anticlockwise directions (Figure 6.8).

Following the design of experiment as per Table 6.4, FSW was processed with nine sets of experiment based upon a Taguchi L9 orthogonal array. Three levels of rotational speed – 1000, 1200 and 1400 rpm; three

Figure 6.8 Experimental setup of FSW process.

Figure 6.9 Welded parts as per L9 orthogonal array.

levels of transverse speed – 30, 40 and 50 mm/min; and three levels of pin diameter – 7, 8 and 9 were varied to perform these sets of experimentation. Figure 6.9 shows the welded parts of ABS-15Al sheets with applying a semi-consumable tool of PA6-50Al.

6.3.8   Sustainability evaluation of welded pieces

For evaluation of mechanical and metallurgical properties, a universal tensile tester, Shore D hardness and inverted microscope were used. Tool consumption rate (TCR) measures a mass of the pin consumed/deposited on the welded parts. The tool consumption rate measures as the welding process of 30 mm of width, so the unit of TCR was fixed as mg/welding of 30 mm. The TCR was measured as:

$$TCR = x_1 - x_2$$

where:
x_1 is the weight of the pin profile before welding
x_2 is the weight of the pin profile after the welding is being processed

Shore D hardness (as per ASTM D2240) is one of the mechanical properties that was measured by pressing a punching knob over the specimen. The stylus configured with an indenter was used to sense the hardness of the materials. The operating process of the Shore D hardness tester is very simple – the pressing of the punching knob until the result is stable is the basic procedure to perform hardness testing.

Following ASTM D2240 for Shore D hardness testing, the Shore D hardness was evaluated under the following expression:

$$S_D = 100 - \frac{20\left(-78.188 + \sqrt{6113.36 + 781.88E}\right)}{E}$$

where:

S_D is the Shore D hardness
E is the elastic modulus of the materials

Tensile properties have been subcategorized as peak strength, peak load, break strength break load, percentage elongation and Young's modulus. Peak strength is the maximum strength of the tensile test specimen before fracture. It was experienced for all specimens tested that fracture strength was lower than the peak strength. The load applied at the peak strength was termed as peak load. A universal tensile tester (as per ASTM D638) was used to calculate the peak load and peak strength of the joints as the function of tensile properties.

Percentage porosity at joints is the morphological property that is usually used to check the joining characteristics. In the welding practices it is usually tried to be kept as low as possible for porosities of joints. The porosity can be the measure of mechanical strengths and is reversible to the mechanical properties because porous structures generally lead to lesser mechanical strength. An inverted microscope with software called metallurgical image analysis software (MIAS) was used to evaluate the morphology of the joints in the form of percentage of porosity. Table 6.5 shows the evaluated mechanical and metallurgical properties of welded pieces.

6.3.9 Optimization of input process variables

Towards optimizing the input process variable for selection of the best contributing process parameters, the variances over signal-to-noise (SN) have been calculated. The SN ratio is always desired to be maximum; conversion of material properties to SN ratio is predicted as either 'smaller is better' or 'larger is better'. For mechanical properties such as peak load, peak strength and Shore D hardness, it is always required to be maximum. For such properties, the SN ratio can be calculated as: 1

$$\eta = -10 \log\left[\frac{1}{n}\sum_{k=1}^{n}\frac{1}{y^2}\right]$$

Table 6.5 Mechanical, tribological and morphological sustainability of welded pieces

Exp no.	Peak load (Kgf)	Break load (KgF)	TCR (mg/30 mm welding)	Peak strength (MPa)	Break strength (MPa)	Peak elongation (mm)	Break elongation (mm)	Shore D hardness	Porosity at joints	Grain size at joint
1	3.41	3.07	14	2.84	2.55	1.90	2.09	77.00	36.48	3.50
2	7.58	6.82	15	6.35	5.71	0.47	2.66	78.50	30.65	3.75
3	2.51	2.26	12	2.09	1.88	1.33	1.33	76.50	42.21	3.50
4	10.35	9.31	21	8.43	7.59	2.47	2.66	80.00	28.06	1.50
5	12.41	11.17	21	10.34	9.31	2.28	2.28	81.50	20.08	1.75
6	9.08	8.17	20	7.57	6.81	2.47	2.66	79.50	33.98	3.00
7	4.99	4.49	14	4.06	3.65	2.85	2.85	79.00	36.65	1.25
8	8.15	7.33	16	6.79	6.11	3.80	3.99	80.00	31.41	1.75
9	4.34	3.90	15	3.53	3.18	3.80	3.80	78.50	42.58	1.75

For properties that desired smaller is better, SN ratios can be calculated as:

$$\eta = -10 \log \left[\frac{1}{n} \sum_{k=1}^{n} y^2 \right]$$

where:

η is the SN ratio

n is the number of experiment

y is the material properties at experiment no. k

Table 6.6 shows the SN ratios for the evaluated properties.

Figure 6.10 shows the SN ratio plot for different output parameters of welded pieces.

Optimizing the process parameter for specific properties, the analysis of variance (ANOVA) table has been drawn by relating the variation of SN ratios over input. Table 6.7 shows the ANOVA table for peak load as calculation of probability (P) and Fishers values. The P value came as <0.05 for rotational speed and proves the significance of the processes (contributed 60.94%).

Peak load was attained more as more of the strength values so that the 'larger is better case' was considered for calculating the SN ratios. Upon this basis, rotational speed ranked 1, transverse speed was ranked 2 and pin diameter ranked 3. Table 6.8 shows the ranking of input process variable over calculating SN ratios variation.

The optimum value of peak load in this case can be predicted by using the following equation:

$$\eta_{opt} = m + \left(m_{A2} - m \right) + \left(m_{B2} - m \right) + \left(m_{C2} - m \right) \tag{6.1}$$

where:

'm' is the overall mean of SN ratio

m_{A2} is the mean of SN ratio for rotational speed at level 2

m_{B2} is the mean of SN ratio for transverse speed at level 2

m_{C2} is the mean of SN data for pin diameter at level 2

Now, for lesser is better by type case

$$y_{opt}^2 = (1/10)^{\eta_{opt}/10} \tag{6.2}$$

Table 6.6 SN ratio (SNR) for properties at different experimental conditions

Exp no.	SNR of peak load	SNR of break load	SNR of TCR	SNR of peak strength	SNR of break strength	SNR of peak elongation	SNR of break elongation	SNR of shore D hardness	SNR of porosity at joints	SNR of grain size at joint
1	10.65	9.73	22.92	9.06	8.15	5.57	6.40	37.72	−31.24	3.50
2	17.59	16.67	23.52	16.05	15.14	−6.55	8.49	37.89	−29.72	3.75
3	7.99	7.07	21.58	6.40	5.48	2.49	2.47	37.67	−32.50	3.50
4	20.29	19.38	26.44	18.51	17.60	7.85	8.49	38.06	−28.96	1.50
5	21.87	20.96	26.44	20.29	19.37	7.15	7.15	38.22	−26.05	1.75
6	19.16	18.24	26.02	17.58	16.66	7.85	8.49	38.01	−30.62	3.00
7	13.96	13.04	22.92	12.17	11.25	9.09	9.09	37.95	−31.28	1.25
8	18.22	17.30	24.08	16.63	15.72	11.59	12.01	38.06	−29.94	1.75
9	12.74	11.83	23.52	10.95	10.04	11.59	11.59	37.89	−32.58	1.75

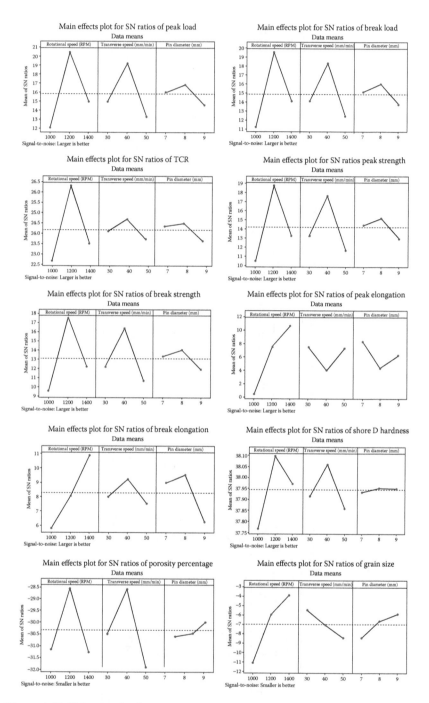

Figure 6.10 SN ratio plot for output parameters.

Table 6.7 Analysis of variance for SN ratios for peak load

Source	DF	Seq SS	Adj SS	Adj MS	F	P	Percentage contribution
Rotational speed (rpm)	2	108.252	108.252	54.126	19.99	0.048	60.94
Transverse speed (mm/min)	2	56.079	56.079	28.040	10.36	0.088	31.57
Pin diameter (mm)	2	7.875	7.875	3.938	1.45	0.407	4.48
Residual error	2	5.415	5.415	2.708			3.04
Total	8	177.622					

Table 6.8 Response table for signal-to-noise ratios (larger is better)

Level speed	Rotational speed (rpm)	Transverse speed (mm/min)	Pin diameter (mm)
1	12.08	14.97	16.01
2	20.45	19.23	16.88
3	14.98	13.30	14.61
Delta	8.36	5.93	2.27
Rank	1	2	3

for properties, larger is better type case

$$y_{opt}^2 = (10)^{n_{opt}/10} \tag{6.3}$$

Calculation:
Overall mean of SN ratio (m) was taken from Minitab software.

$$m = 15.83 \text{ dB (Table 6.6)}$$

Now, from response table of signal-to-noise ratio, $m_{A2} = 20.45$, $m_{B2} = 19.23$, $m_{C2} = 16.88$ (from Table 6.8)

Now, taking Equation (6.1), putting the values

$$\eta_{opt} = 15.83 + (20.45 - 15.83) + (19.23 - 15.83) + (16.88 - 15.83)$$

$$\eta_{opt} = 24.90$$

Now, from equation (6.3) $y_{opt}^2 = (10)^{nopt/10}$

$$y_{opt}^2 = (10)^{24.90/10}$$

$$y_{opt} = 17.58 \text{ N}$$

The predicted optimum value for peak load = 17.58 N.

By applying the aforementioned equation predicted value, it can be determined towards optimisation of process parameters for the remaining output parameters (mechanical properties and metallurgical properties). Based upon Tables 6.6 through 6.8, the predicted and experimentally determined values of output parameters (mechanical and metallurgical properties) and actual values have been shown in Table 6.9.

6.3.9.1 Multifactor optimization of input process variable

As SN ratios of all the properties have been combined by selecting SN ratios on a larger is better basis, 1200 rpm, 40 mm/min transverse speed and 7 mm pin diameter resulted as the best setting and were further selected as the best set of input process parameters (Figure 6.11).

Since P values of RPM and transverse speed were obtained as 0.002 and 0.006, which is less than 0.05, the combined optimized setup was predicted as significant. RPM and transverse speed appeared to be the most significant contributing factors as 57.38% and 41.11% contributing percentage (Table 6.10).

As obtained from the response in Table 6.11, the rotational speed was ranked as 1, transverse speed as 2, and pin diameter was achieved as 3.

A stirred zone of welded pieces was being examined through an optical microscope by support of software known as 'metallurgical image analysis software (MIAS)'. All the welded pieces were examined at 100× of magnification level, and porosity was measured at that magnification level. The welded joints processed at 1000 and 1400 rpm contained some porous significance on their surfaces whereas welded parts obtained at 1200 rpm appeared as less porous structures and observed a very stirred surface with excellent metal powder polymer mixing (Figure 6.12). In relation with mechanical properties to the porosity of the stirred joints, mechanical properties were varied accordingly with porosity. At experiment no. 5, where mechanical properties resulted in maximum value, porosity came as a minimum value, which justified the reason for the joints' strength. Improper mixing of the joints at experiment no. 9 resulted in maximum porosities where porosity can be observed – this resulted in minimum mechanical properties, a proven reason for strength variation.

Table 6.9 Predicted versus actual values of output parameter at optimum predicted setting

Properties	Peak load (Kgf)	Break load (KgF)	TCR (mg/30 mm welding)	Peak strength (MPa)	Break strength (MPa)	Peak elongation (mm)	Break elongation (mm)	Shore D hardness	Porosity at joints	Grain size at joint
Predicted value	17.58	15.85	22.80	14.58	13.09	5.02	4.54	81.56	21.47	1.33
Value at predicted setting	17.41	15.56	21.87	14.34	13.21	4.95	4.61	81.5	21.08	1.50

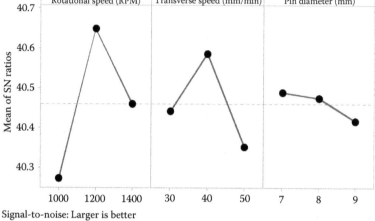

Signal-to-noise: Larger is better

Figure 6.11 Linear model for combined SN ratio.

Table 6.10 Analysis of variance for SN ratios

Source	DF	Seq SS	Adj SS	Adj MS	F	P	Percentage contribution
Rotational speed (rpm)	2	0.214349	0.214349	0.107174	450.34	0.002	69.73
Transverse speed (mm/min)	2	0.083962	0.083962	0.041981	176.40	0.006	27.30
Pin diameter (mm)	2	0.008566	0.008566	0.004283	18.00	0.053	2.78
Residual error	2	0.000476	0.000476	0.000238			0.15
Total	8	0.307353					

Table 6.11 Response table for signal-to-noise ratios (larger is better)

Level	Rotational speed (rpm)	Transverse speed (mm/min)	Pin diameter (mm)
1	40.27	40.44	40.49
2	40.65	40.59	40.47
3	40.46	40.35	40.42
Delta	0.38	0.23	0.07
Rank	1	2	3

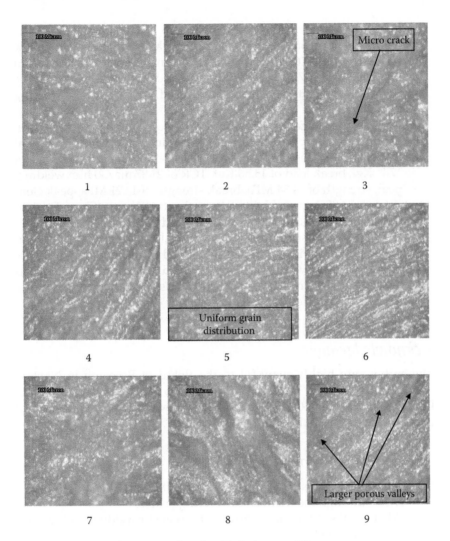

Figure 6.12 Optical micrographs of welded pieces at 100×.

6.4 Summary

- Welding of two virgin dissimilar thermoplastics are impractical as they exhibit different molecular, melting, rheology, thermal and mechanical properties. In the present case, reinforcement of Al metal powder in a thermoplastic matrix offered the similar range of MFI, which helped achieve successful welding of dissimilar thermoplastic materials. Reinforcement of 15% Al content to ABS and 50% Al content to a PA6 matrix resulted in the similar MFI range of 11.57 g/10 min and 11.97 g/10 min, respectively.

- Reinforcement of 15% Al content in ABS led to an increase in the melting point from 201.22°C to 218.11°C, and reinforcement of 50% Al content in PA6 led to modification in the melting point from 218.35°C to 218.27°C. The reinforcement caused the material to be thermally compatible to each other used for friction stir welding application.
- With the optimization of the input process variable, the best predicted parametric setting has been obtained by using a statistical model. At the best setting, the values obtained are peak load of 17.41 Kgf, break load of 15.56 Kgf, TCR of 21.87 mg/30 mm welding, peak strength of 14.34 MPa, break strength of 13.21 MPa, peak elongation of 4.95 mm and break elongation of 4.61 mm. Shore D hardness of 81.5, porosity at joint of 21.08% and grain size of 1.50.
- ANOVA results suggested that rotational speed was the most significant contributing factor as it provided 69.73% for changes in SN ratios. The transverse speed was ranked 2 and pin diameter was ranked 3. The pin diameter was the least significant contributing factor as it provided 2.78% only.

Acknowledgement

The authors are highly thankful to the Board of Research in Nuclear Science (BRNS) No: 34/14/10/2016-BRNS/34036 and the University Grants Commission (UGC) F.30-66/2016 (SA-II) for providing financial assistance to carry out the research work.

Theoretical questions

1. What is friction stir welding? Describe the process of joining two sheets through the friction stir welding process.
2. What are the differences between friction stir welding and friction stir processing?
3. The advancing side in friction stir welding/processing can be considered a side where flow of materials has a higher temperature whereas the retreading side is the phase where material flow has a lower temperature. (T/F)
4. Friction stir welding is a liquid state welding technique. The process carried above the melting point of substrate. (T/F)
5. This FSP technique is not considerable for the joining purposes; friction stir processing is applicable to only where refinement of microstructures is required. (T/F)
6. FSW and FSP can be differentiated by investigating advancing and _____ side.

7. The shoulder arrangement creates the frictional heating, and a pin probe facilitates the _____ to the joint interface.
8. Choose the correct answer:
    i. Which instrument is used to evaluate the melting point/ solidification point/glass transition temperature of thermoplastic material?
        a. DSC
        b. MFI tester
        c. 3D printer
        d. Twin screw extruder
    ii. Which is a solid-state welding process?
        a. Submerged arc welding
        b. Gas metal arc welding
        c. Tungsten inert gas welding
        d. Friction stir welding
    iii. How many experiments are created by a Taguchi L9 orthogonal?
        a. 3
        b. 9
        c. 18
        d. 16
9. What is melt flow index? Which MFI ASTM standard is used for most of the thermoplastic material? How can the melt flow index of any thermoplastic materials be modified?
10. If molecular weight of a thermoplastic material is changed, then what is the effect of the melt flow index of that thermoplastic material?
11. What are the process variables of a twin screw extrusion and fused deposition modelling setup? State any three of each.
12. What is shore D hardness? State the ASTM standard used for a shore D hardness tester. Give the mathematical equation to determine shore D hardness in terms of Young's modulus.

Numerical questions (solved)

1. If Young's modulus of a thermoplastic material is evaluated as 0.1 MPa by the tensile tester, what is the hardness of that thermoplastic material by a Shore D hardness tester?
   Solution:
   By considering the relation of Shore D hardness in terms of Young's modulus as:

$$S_D = 100 - \frac{20\left(-78.188 + \sqrt{6113.36 + 781.88E}\right)}{E}$$

where:

S_D is the Shore D hardness

E is Young's modulus

Taking E = 100 Mpa in the equation

$$S_D = 100 - \frac{20\left(-78.188 + \sqrt{6113.36 + 781.88 \times 100}\right)}{100}$$

$$S_D = 100 - 42.43$$

$$S_D = 57.57 \text{ shore D}$$

2. The peak strength of a welded piece was obtained as 0.5428 kg/mm². What would be the signal-to-noise ratio for peak strength at that experimental setup, considering the 'larger is better case' and 'smaller is better case'?

Solution:

Considering the equation of signal-to-noise ratio for 'larger is better case'

$$\eta = -10 \log \left[\frac{1}{n} \sum_{k=1}^{n} \frac{1}{y^2} \right]$$

Here η is signal-to-noise ratio, n = 1 as number of experiment, y = 0.5428

So, putting the values:

$$\eta = -5.30669 \text{ dB}$$

Now, considering smaller is better case:

$$\eta = -10 \log \left[\frac{1}{n} \sum_{k=1}^{n} y^2 \right]$$

$$\eta = 5.30669 \text{ dB}$$

3. For predicting the optimized value of a property of welded piece, if the η_{opt} was calculated 27.80, then what would be the predicted value of property? Consider both the cases – that is, larger is better and smaller is better.

Solution:
For properties 'smaller is better' case

$$y_{opt}^2 = (1/10)^{n_{opt}/10} \tag{iii}$$

For properties 'larger is better' type case

$$y_{opt}^2 = (10)^{n_{opt}/10} \tag{iv}$$

Considering 'larger is better' case

$$y_{opt}^2 = (10)^{n_{opt}/10}$$

$$y_{opt}^2 = (10)^{27.80/10}$$

$$y_{opt}^2 = (10)^{2.78}$$

$$y_{opt}^2 = 602.5595$$

$$y_{opt} = 24.54$$

Considering 'smaller is better' case

$$y_{opt}^2 = (1/10)^{n_{opt}/10}$$

$$y_{opt}^2 = (1/10)^{27.80/10}$$

$$y_{opt}^2 = (1/10)^{2.78}$$

$$y_{opt}^2 = 0.00166$$

$$y_{opt} = 0.0407$$

4. As predicting the optimized value of Young's modulus (MPa), the value of overall mean = 47.70 dB and the optimum equation was given as; $n_{opt} = m + (m_{A2} - m) + (m_{B3} - m) + (m_{C1} - m)$, where $m_{A2} = 48.04$, $m_{B3} = 47.86$, $m_{C1} = 48.86$. Calculate the predicted value as optimized by the statistical model. Consider 'larger is better' case.
Solution:
For optimisation, the following formula has been used:

$$n_{opt} = m + (m_{A2} - m) + (m_{B3} - m) + (m_{C1} - m)$$

where:
    'm' is the overall mean of S/N data
    m_{A2} is the mean of S/N data for A at level 2
    m_{B3} is the mean of S/N data for B at level 3
    m_{C1} is the mean of S/N data for C at level 1

$$y_{opt}^2 = (1/10)^{n_{opt}/10} \qquad \text{for properties, lesser is better}$$

$$y_{opt}^2 = (10)^{n_{opt}/10} \qquad \text{for properties, larger is better}$$

Calculation:
    Overall mean of SN ratio (m) was taken from Minitab software.

$$m = 47.70$$

Now, from response table of signal-to-noise ratio, $m_{A2} = 48.04$, $m_{B3} = 47.86$, $m_{C1} = 48.86$.
    From here,

$$\eta_{opt} = 47.70 + (48.04 - 47.70) + (47.86 - 47.70) + (48.86 - 47.70)$$

$$\eta_{opt} = 49.36 \text{ db}$$

Now,

$$y_{opt}^2 = (10)^{n_{opt}/10}$$

$$y_{opt}^2 = (10)^{49.36/10}$$

$$y_{opt} = 293.77 \text{ MPa}$$

References

1. R. Singh, R. Kumar, I. Mascolo and M. Modano, On the applicability of composite PA6-TiO_2 filaments for the rapid prototyping of innovative materials and structures, *Compos. Part B Eng.* 143 (2018), 132–140.
2. R. Kumar, R. Singh, D. Hui, L. Feo and F. Fraternali, Graphene as biomedical sensing element: State of art review and potential engineering applications, *Compos. Part B Eng.* 134 (2018), 193–206.
3. R. Singh, R. Kumar, L. Feo and F. Fraternali, Friction welding of dissimilar plastic/polymer materials with metal powder reinforcement for engineering applications, *Compos. Part B Eng.* 101 (2016), 77–86.
4. R. Kumar, R. Singh, I.P.S. Ahuja, R. Penna and L. Feo, Weldability of thermoplastic materials for friction stir welding—A state of art review and future applications, *Compos. Part B Eng.* 137 (2018), 1–15.

5. R. Kumar, R. Singh, I.P.S. Ahuja, A. Amendola and R. Penna, Friction welding for the manufacturing of PA6 and ABS structures reinforced with Fe particles, *Compos. Part B Eng.* 132 (2018), 244–257.

6. R. Singh, R. Kumar, N. Ranjan, R. Penna and F. Fraternali, On the recyclability of polyamide for sustainable composite structures in civil engineering, *Compos. Struct.* 184 (2018), 704–713.

7. R. Singh, R. Kumar and N. Ranjan, Sustainability of recycled ABS and PA6 by banana fiber reinforcement: Thermal, mechanical and morphological properties, *J. Inst. Eng. Ser. C* (2018), 1–10. doi:10.1007/s40032-017-0435-1.

8. R. Kumar, R. Singh and I.P.S. Ahuja, Investigations of mechanical, thermal and morphological properties of FDM fabricated parts for friction welding applications, *Meas. J. Int. Meas. Confed.* 120 (2018), 11–20.

9. R. Kumar and R. Singh, Prospect of graphene for use as sensors in miniaturized and biomedical sensing devices, in S. Hashmi (Ed.), *Reference Module in Materials Science and Materials Engineering*, pp. 1–13. Oxford, UK: Elsevier, 2018.

10. R. Singh, R. Kumar and I.S. Ahuja, Thermal analysis for joining of dissimilar polymeric materials through friction stir welding, in S. Hashmi (Ed.), *Reference Module in Materials Science and Materials Engineering*. Oxford, UK: Elsevier, 2017.

11. R. Singh and R. Kumar, Development of low-cost graphene-polymer blended in-house filament for fused deposition modeling, in S. Hashmi (Ed.), *Reference Module in Materials Science and Materials Engineering*. Oxford, UK: Elsevier, 2017.

12. R. Singh, R. Kumar and M.S.J. Hashmi, Friction welding of dissimilar plastic-based material by metal powder reinforcement, in S. Hashmi (Ed.), *Reference Module in Materials Science and Materials Engineering*, Vol. 13. Oxford, UK: Elsevier, 2017.

13. R. Singh, R. Kumar and S. Kumar, Polymer waste as fused deposition modeling feed stock filament for industrial applications, in S. Hashmi (Ed.), *References Module in Materials Science and Materials Engineering*, pp. 1–12. Oxford, UK: Elsevier, 2017.

14. E. Azarsa and A. Mostafapour, On the feasibility of producing polymer-metal composites via novel variant of friction stir processing, *J. Manuf. Process.* 15 (2013), 682–688.

15. E.A. Squeo, G. Bruno, A. Guglielmotti and F. Quadrini, Friction stir welding of polyethylene sheets, 2009, pp. 241–146.

16. http://proceedings.asmedigitalcollection.asme.org/, accessed 31 January 2016. Terms of Use: http://www.asme.org/about-asme/terms-of-use, 2016, pp. 1–4.

17. P.H.F. Oliveira, S.T. Amancio-Filho, J.F. Santos and E.H. Jr., Preliminary study on the feasibility of friction spot welding in PMMA, *Mater. Lett.* 64 (2010), 2098–2101.

18. F. Simões and D.M. Rodrigues, Material flow and thermo-mechanical conditions during Friction Stir Welding of polymers: Literature review, experimental results and empirical analysis, *Mater. Des.* 59 (2014), 344–351.

19. A.B. Abibe, M. Sônego, J.F. dos Santos, L.B. Canto and S.T. Amancio-Filho, On the feasibility of a friction-based staking joining method for polymer-metal hybrid structures, *Mater. Des.* 92 (2016), 632–642.

20. J. Gonçalves, J.F. Dos Santos, L.B. Canto and S.T. Amancio-Filho, Friction spot welding of carbon fiber-reinforced polyamide 66 laminate, *Mater. Lett.* 159 (2015), 506–509.

21. G. Buffa, D. Baffari, D. Campanella and L. Fratini, An innovative friction stir welding based technique to produce dissimilar light alloys to thermoplastic matrix composite joints, *Procedia Manuf.* 5 (2016), 319–331.

22. A. Paoletti, F. Lambiase and A. Di Ilio, Optimization of friction stir welding of thermoplastics, *Procedia CIRP* 33 (2015), 563–568.

23. B. Vijendra and A. Sharma, Induction heated tool assisted friction-stir welding (i-FSW): A novel hybrid process for joining of thermoplastics, *J. Manuf. Process.* 20 (2015), 234–244.

24. M. Rezaee Hajideh, M. Farahani, S.A.D. Alavi and N. Molla Ramezani, Investigation on the effects of tool geometry on the microstructure and the mechanical properties of dissimilar friction stir welded polyethylene and polypropylene sheets, *J. Manuf. Process.* 26 (2017), 269–279.

25. A. Hamdollahzadeh, M. Bahrami, M.F. Nikoo, A. Yusefi, M.K.B. Givi and N. Parvin, Microstructure evolutions and mechanical properties of nano-SiC-fortified AA7075 friction stir weldment: The role of second pass processing, *J. Manuf. Process.* 20 (2015), 367–373.

26. D. Gamit, R.R. Mishra and A.K. Sharma, Joining of mild steel pipes using microwave hybrid heating at 2.45 GHz and joint characterization, *J. Manuf. Process.* 27 (2017), 158–168.

27. S. Eslami, T. Ramos, P.J. Tavares and P.M.G.P. Moreira, Shoulder design developments for FSW lap joints of dissimilar polymers, *J. Manuf. Process.* 20 (2015), 15–23.

28. F. Lambiase, A. Paoletti, V. Grossi and A. Di Ilio, Friction assisted joining of aluminum and PVC sheets, *J. Manuf. Process.* 29 (2017), 221–231.

29. E. Azarsa and A. Mostafapour, Experimental investigation on flexural behavior of friction stir welded high density polyethylene sheets, *J. Manuf. Process.* 16 (2014), 149–155.

30. R.S. Mishra and Z.Y. Ma, Friction stir welding and processing, *Mater. Sci. Eng. R Rep.* 50 (2005), 1–78.

31. V. Sharma, U. Prakash and B.V.M. Kumar, Surface composites by friction stir processing: A review, *J. Mater. Process. Technol.* 224 (2015), 117–134.

32. H.K. Mohanty, M.M. Mahapatra, P. Kumar and P.K. Jha, Modeling the effects of tool geometries on the temperature distributions and material flow of friction stir aluminum welds, in L. Nastac, L. Zhang, B.G. Thomas, A. Sabau, N. El-Kaddah, A.C. Powell, and H. Combeau (Eds.), *CFD Modeling and Simulation in Materials Processing*, pp. 25–32. Hoboken, NJ: John Wiley & Sons, 2012.

chapter seven

A novel approach of using ultrasound to improve the surface quality of 3D printed parts

S. Maidin, E. Pei and M. K. Muhamad

Contents

7.1 Background: Surface finishing of 3D printed parts

The surface finish of products is an important factor for users and manufacturers for function and safety. The quality of the surface finish relates to the perceived appearance and its aesthetic value. In conventional manufacturing, the product surface finish can be improved through methods such as spray painting, sandblasting, polishing and anodizing. As the use of Additive Manufacturing (AM) grows, there is an expectation that end-user part production will become more mainstream [1]. AM technologies allow users to build functional components with complex geometrical shapes within a reasonable build time. Fused Deposition Modelling (FDM) is an AM method where predetermined layers of extruded material are deposited in a controlled way to create a solid three-dimensional part. Although the International Organization for Standardization Standard Practice Guide for Design for Additive Manufacturing [2] defines this process as material extrusion where material is selectively dispensed through a nozzle [2], the term Fused Deposition Modelling (a trademark of Stratasys, Ltd.) will be used for this paper as it is more commonly recognized for this technology.

FDM systems are popular due to the availability of a wide range of materials in a filament form. Their superior mechanical properties, versatility, low cost and wide use in rapid prototyping also make FDM systems an ideal method of producing parts for a number of applications [3]. In the FDM process, the material is melted and extruded via a heated nozzle to form successive layers that build up the cross section of parts. The three-dimensional part takes the form of vertically stacked layers consisting of contiguous material fibres (rasters) with interstitial voids (air gaps) [4]. The main disadvantage of the FDM process is that seam lines appear between each layer and excess material is sometimes produced as a residue, leading to surface roughness and poor finish [5]. A thicker seam line produces more visibly evident stair-stepping, which consequentially contributes to a rougher surface finish. A thinner seam line can be achieved by having a thinner layer thickness, but doing this increases the duration of the print time.

Bual and Kumar [6] proposed that improving the surface finish of FDM parts can be divided into four categories: optimizing the build orientation; slicing strategy (layer thickness); optimizing the fabrication parameters; and through post-treatment. In the first category, using a proper orientation for printing can lead to a better surface finish, and researchers have developed techniques such as the use of algorithms to determine the optimum orientation based on the user's selection of primary criteria and the optimal thickness of the layers [7–9]. The work by [10] also used algorithm-based techniques to find the optimum orientation of the CAD model to minimize the build time, reducing staircase error and minimizing material to improve the part quality and economy during manufacture. Second, in terms of slicing strategy, a thinner slice layer produces better surface finish but it consequentially increases the build time. Researchers have found that a better surface quality can be realized by producing parts with a thinner layer, and this is generally considered to be the most accepted method to achieve good finishing [11,12]. Third, the surface finish depends on a number of process parameters of the FDM machine and includes the layer thickness, raster angle, air gap, contour width, contour depth, part raster width and good stacking sequence of the vertically bonded layers [13]. For example, the larger the contour width, the better the profile accuracy, while a narrow width results in a dense and good appearance of the aperture area on the surface layer. Lastly, there have been various studies related to post treatment of FDM parts to improve the surface finish. In terms of post processing, some common methods include manually trimming the support structures, sanding, tumbling, barrel finishing, machining, sandblasting, polishing, heat treating and applying coating to achieve the desired surface qualities or properties. Other researchers proposed the use of CNC machining [14,15], which incorporated a virtual model offset to be generated in the

Computer-Aided Manufacturing of the machine code. Another approach is the use of solvent-based chemicals such as acetone that partially dissolve the surface of the material, resulting in a smoother surface finish [16,17]; or the use of a solution of dimethyl ketone to treat the external filaments without affect object shape [3]. Rao et al. [18] analysed various parameters of chemical treatment processes including concentration and exposure time. They used two different chemicals, dimethyl ketone and methyl ethyl ketone. For the dimethyl ketone solution, using the optimal concentration was vital, whereas for methyl ethyl ketone, temperature and time were more significant parameters. However, these techniques expose the operator to toxic fumes and chemicals, and it is time consuming and costly to undertake. If not well controlled, there could be unintentional surface modifications made to the 3D printed parts because they are based on mechanical and chemical actions in a manual or semiautomatic way [19]. It is therefore important to consider the time and expertise involved in post-processing activities for 3D printed parts. If the volume and the number parts are large, then the use of automated post-processing methods should be considered, such as the hybrid rapid prototyping system proposed by [15], which incorporates layer-by-layer machining and deposition to achieve improved surface finish and functionality of AM parts.

7.2 The use of ultrasound to improve the surface finish of parts

The use of ultrasound has become increasingly used in many industrial applications, and it is a proven technology that can improve the quality of a machined surface finish [5,20–22]. Ultrasonic piezoelectric transducers are capable of converting electrical signals into mechanical energy, in the form of sound and vibration. Ultrasonic vibrations with piezoelectric components that pulse in a vertical direction have been used to assist laser machining, and it has produced results with a better surface finish [23]. Tabatabaei et al. [24] reported that ultrasonic assisted machining is an advanced processing technology that has the capability to improve the machining process, especially for use on harder materials. In ultrasonic machining, the tool vibrates at a high frequency, usually more than 20 kHz, and abrasive slurry is pumped between the workpiece and the tool [25]. This process does not cause a chemical reaction and is therefore regarded to be safe. More importantly, this process does not chemically corrode the workpiece. Friel and Harris [26] studied the use of ultrasonic AM in a hybrid production process and found that these processes are also suited for high-end metal matrix composites at high temperatures and pressures. Nik et al. [27] also studied the effect of ultrasonic grinding of Ti6A14v alloys at a frequency range of 20 kHz applied onto a work-piece, and the results show an evident improvement of surface roughness. Taking a step further, Janaki Ram et al. [28] proposed a novel method of Ultrasonic Consolidation (UC), which is a direct AM process developed for fabrication of metallic parts from foils using Al alloy 3003 foils as the base material together with a number of engineering materials such as Ni-base alloy Inconel and stainless steel. Despite these benefits, wide application of this technology is still limited. This is due to the high cost of designing an ultrasonic system and the cost associated with energy that is required to induce the vibrations [29].

Piezoelectric transducers used for ultrasonic assisted machining (UAM) generate mechanical motion through the piezoelectric effect through the use of certain materials, such as quartz or lead zirconate titanate. In contrast, magnetostrictive transducers are usually constructed from a laminated stack of nickel or nickel alloy sheets [30]. Applying ultrasonic frequencies to the FDM process has the potential to improve the surface finish of parts, and this research proposes the use of piezoelectric ultrasonic transducers by converting low-frequency electrical energy (60 Hz) to a high-frequency electrical signal (approximately 20 kHz) that is fed to the transducer component. There is also a potential for ultrasonic-based finishing achieving good surface results across the component without being constrained by the geometrical profile of the part. For example, recent work on surface finishing for FDM parts by Boschetto and Bottini [19] who suggested the use of Barrel Finishing (BF) pointed out that their technique is closely influenced by the deposition angle of the FDM surface and that their measurements showed different roughness decreasing with BF working time correspondent to different surface slopes.

7.3   Methodology

The aim of this study is to investigate the possibility of applying ultrasonic vibrations in the FDM process and to investigate their influence on the surface finish of parts. It is anticipated that the use of ultrasonic technology can overcome some of the difficulties brought by post-processing of FDM parts. The main objectives of the research are to investigate the feasibility of using ultrasonic transducers to improve the surface finish of a 3D printed part; and to determine the results in terms of different frequencies of oscillations produced by the ultrasonic-assisted FDM system. Three different setups were developed, incorporating the use of an ultrasonic transducer installed onto an FDM system. The UP Plus 2 FDM printer was chosen due to its popularity among users, as well as its availability. It has a build size of 140 × 140 × 135 mm and a 0.4 mm nozzle diameter was used to achieve fine build parameters through its thin layer thickness and speed. The open nature of the printer allows slight modifications to be made to achieve a fully integrated ultrasonic-assisted system. The use of Acrylonitrile Butadiene Styrene (ABS) material was selected due to its cost-effectiveness and availability. The three setups summarized in Table 7.1 were evaluated according to the ease of installation, the level of safety and the amount of reliability. Concept 1 (Figure 7.1) was the most radical, as the ultrasonic transducer would be integrated with the heating block rather than the nozzle. Although this would avoid affecting the extrusion process, the main concern was whether the high temperature of the heated block could cause the transducer's performance to drop. Tests would also need to be carried out to

Table 7.1 Concepts of an ultrasonic-assisted FDM system

	Concept 1	Concept 2	Concept 3
Description of setup	The transducer is directly fixed onto the heated block.	The transducer is mounted on a supporting plate, which is attached to the nozzle.	The transducer is fastened onto the build platform or the workpiece.
Installation	A fastening device will be soldered onto the heated block and secure the transducer in place.	Non-heat conducting sheet metal will be used to hold the transducer that is welded onto the nozzle.	The transducer is attached onto the side of the build platform or workpiece to enable direct vibration.
Safety	The performance of the transducer should be monitored as the heat could cause the transducer's performance to drop.	The material used for the mounting plate should not conduct heat. The length of the plate should be as short as possible so that energy loss from the vibration can be reduced.	The location of the transducer must not affect the movement of the nozzle or the build process.
Reliability	The transducer mounted onto the heating block should not inhibit the flow of the material or the movement of the nozzle.	The mounting plate should be securely fastened and not affect the movement of the nozzle or the build process.	Too much vibration could cause the overall print quality to be affected.

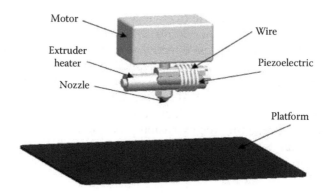

Figure 7.1 Concept 1.

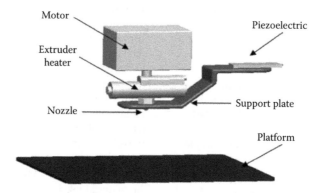

Figure 7.2 Concept 2.

determine whether the heated block and nozzle could withstand constant vibration from the transducer. Concept 2 (Figure 7.2) is a variation of the first concept where the ultrasonic transducer would be placed at a distance from the heated block and the nozzle using a heat insulating plate. The main concern was whether the supporting plate would absorb most of the ultrasonic vibration and render the device ineffective. Concept 3 (Figure 7.3) required the ultrasonic transducer to be clamped onto the build platform to enable direct transmission of vibration to the sample during the build process.

As the nozzle is an essential component in the FDM system, a vibration analysis must first be carried out to ensure that the nozzle is able to withstand the high frequency of vibration and is essential for risk assessment and operator safety. For the initial part of this study, computer-aided engineering (CAE) was used to investigate the components of the nozzle

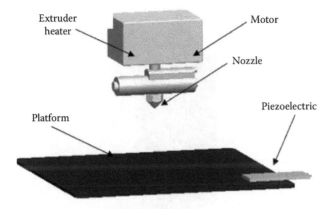

Figure 7.3 Concept 3.

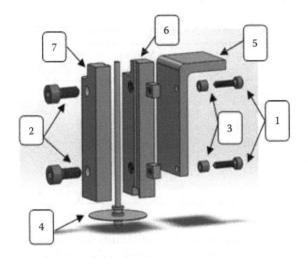

Figure 7.4 Exploded view of FDM model.

to ensure that the mechanical parts would function when subjected to ultrasonic vibration. Solidworks CAD software was used, and Figure 7.4 shows an exploded view of the FDM nozzle component consisting of seven parts made from stainless steel and magnesium alloy. Figure 7.5 shows the assembly view of the FDM nozzle, and Table 7.2 shows a detailed breakdown of each part. Static structural analysis was carried out using ANSYS, which is a finite element analysis tool for structural analysis, including linear, nonlinear and dynamic studies. ANSYS was used to analyse the effect of the nozzle when subjected to ultrasonic vibrations from the transducer. The total deformation of the FDM nozzle, virtual

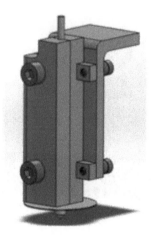

Figure 7.5 Assembly view of FDM nozzle.

Table 7.2 Parts of the FDM nozzle

No	Part	Quantity	Volume (mm³)	Material	Density (kg/m³)
1	Cap screw M3	2	120.54		
2	Cap screw M5	2	413.72	Stainless steel	7750
3	Bushing	1	50.27		
4	Extrusion nozzle	1	444.87		
5	Guider	1	8134.91		
6	Back clamper	1	8728.59		
7	Front clamper	1	8373.06	Magnesium alloy	1800

stress tests and the Factor of Safety (FoS) limits were obtained through the ANSYS simulation. The total deformation of the FDM nozzle and equivalent stress with frequencies ranging from 20 to 30 kHz and 30 to 40 kHz were studied.

Figure 7.6 shows results of the equivalent stress of the FDM nozzle when subjected to a vibration frequency of between 20 and 30 kHz. The highest FoS value that was obtained was 11.721 Mpa, and the lowest value was 21.324 MPa. The calculations showed that the FoS value of the model was 20.56, referring to the fact that the nozzle can withstand a frequency range of between 20 and 30 kHz when subjected to vibration forces from the ultrasound transducer. Even though the FoS was high, we observed that slight bending occurred on the nozzle due to the fact that the nozzle component has a thin thickness profile.

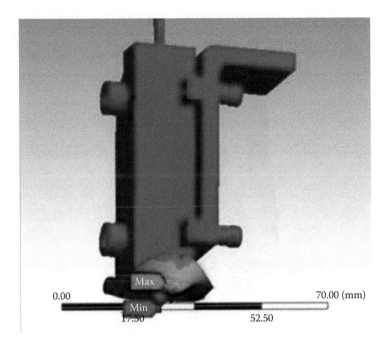

Figure 7.6 Equivalent stress of FDM nozzle using frequency 20–30 kHz.

$$\text{FoS} = \frac{\text{Ultimate tensile strength}}{\text{Maximum stress}}$$

$$= \frac{241\,\text{MPa}}{11.721\,\text{Mpa}}$$

$$= 20.56$$

Next, we subjected the nozzle to a vibration frequency of 30–40 kHz from the ultrasonic transducer. The highest value of maximum stress that was obtained was 12.753 Mpa, and the lowest value was 6.464 MPa. The FoS for this model was 18.8975, which was higher than the 20–30 kHz frequency range. The ultimate tensile strength was higher because the frequency was increased. The supposition is that having a higher frequency will result in a lower FoS for the FDM nozzle. At this frequency range, we observed that high deformations had occurred on the extrusion nozzle (Figure 7.7), making the component potentially unsafe for use.

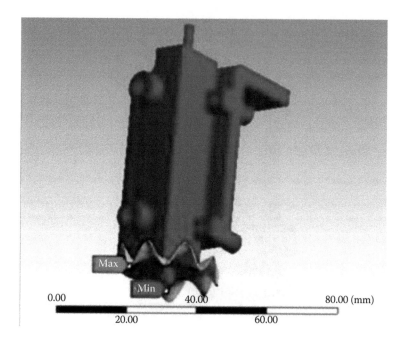

Figure 7.7 Equivalent stress of FDM nozzle using frequency 30–40 kHz.

$$\text{FoS} = \frac{\text{Ultimate tensile strength}}{\text{Maximum stress}}$$

$$= \frac{241\,\text{MPa}}{12.753\,\text{Mpa}}$$

$$= 18.8975$$

Based on the findings of Concept 1, it was decided that this setup would inevitably affect the heated block and the nozzle in the long term. Mousavi et al. [31] undertook work to examine the effects of ultrasonic vibrations on forward extrusion processes and showed that while applying the ultrasonic vibrations had no significant effect on the equivalent plastic strain of the material, the extrusion force fluctuated and reduced when subjected to axial ultrasonic vibrations. Although their work was based on conventional extrusion processes, the overall process is similar to the FDM technique. Due to the fact that Concept 1 had a higher risk of failure, Concept 3 was seen to be safer and could offer positive results.

Furthermore, the process of conceptual design selection is based on the product design specification. All the conceptual designs must be evaluated by a weight decision matrix to select the most appropriate design. The design criteria were ranked with weighted factor and scoring of the

Table 7.3 Weight decision matrix

Design criteria	Weight factor (%)	Conceptual design					
		1		2		3	
		Score	Rating	Score	Rating	Score	Rating
Installation	35	6	2.10	6	2.10	7	2.45
Safety	35	3	1.05	7	2.45	7	2.45
Reliability	30	7	2.10	8	2.40	8	2.40
Total rating	100		5.25		6.95		7.30

design to the criteria. Therefore, all the criteria were ranked according to the product design specification with a different percent of weight based on 100% weight factor. Subsequently, all of the idea concepts were given a score out of 10 marks, and the higher score would be the better criteria of the specific idea concept. The criteria for the selection of the best setup consists of how easy was the installation, how safe was the printer and ultrasonic piezoelectric transducer during printing, and how reliable was the transducer that was mounted to the 3D printer. Table 7.3 illustrates the table of a weight decision matrix. Referring to the total rating of 7.30, the Design Concept 3 showed the ultrasonic piezoelectric transducer being clamped onto the surface of the platform without being exposed to high temperature and having direct transmission of vibration to the sample, which was seen as the safest approach and the most feasible among all three concepts. A decision was made for a prototype to be built and tested and a standard piezoelectric ultrasonic transducer that operated in a horizontal vibration mode was securely mounted onto the build platform of the FDM machine (Figures 7.8 and 7.9). The piezoelectric device that was

Figure 7.8 UP Plus 2 3D printer.

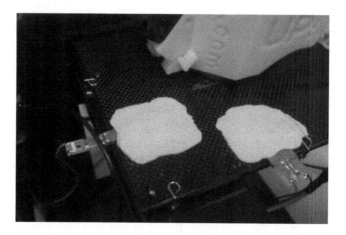

Figure 7.9 Piezoelectric transducer attached to platform.

fastened onto the build platform would enable ultrasonic vibration to be transmitted directly onto the built part with minimal energy loss. The key challenge of this setup was to ensure that the piezoelectric transducer is properly mounted without touching other parts of the 3D printer during the fabrication process. The overall experimental setup is illustrated in Figure 7.10.

To enable control and monitoring of the setup, we used a Function Generator (Figure 7.11) in the form of a Combo Tester ED-770 with the ability to produce a maximum power of 100 W with a 20 kHz frequency. It is an instrument used to generate a variety of synthesised electrical signals and repetitive waveforms over a wide range of frequencies. This equipment had a maximum power of 20 V having an adjustable frequency to power the piezoelectric component (Figure 7.12), which was supplied by Piezo Systems, INC. with a weight of 2.3 g, a deflection of ±3.6 μm and a maximum voltage of ±90 Vp. The amplitude of no-load vibration of the actuator was 10 μm; however, it was not measured under load. To amplify the generated vibration, the transducer had to be stimulated around its natural frequency. The natural frequency of the actuator was measured using an oscilloscope being attached to the transducer as shown in the set up illustrated in Figure 7.10. For the experiments, the frequency was set at 11, 16, 21, 27, 40 and 50 kHz. A total of seven models were printed using the FDM UP Plus 2 3D printer including a control test piece that was fabricated without any ultrasonic vibration being subjected. For the experiments, we used an air gap of semi-solid honeycomb, a layer thickness of 0.20 mm and vibrated by an ultrasonic frequency of 11, 16, 21, 27, 40 and 50 kHz with constant amplitude of 10 μm as the standard frequency.

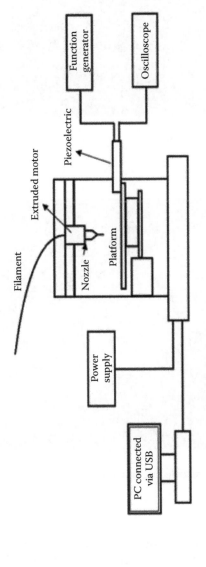

Figure 7.10 Schematic diagram of Concept 3 showing the ultrasonic-assisted FDM system.

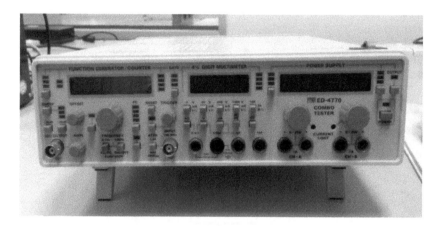

Figure 7.11 Function generator.

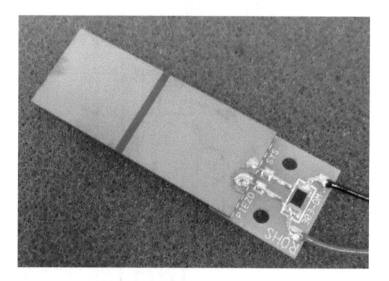

Figure 7.12 Ultrasonic piezoelectric transducer.

A block model measuring 20 × 10 × 15 mm was created in Solidworks and printed using the UP 2 FDM printer (Figure 7.13). Next, a scientific-grade optical microscope with aid of VIS-Pro image analysis and measurement software version 2.90 was used to investigate the layer thickness of the printed samples (Figure 7.14). A Mitutoyo surface

Figure 7.13 Printed sample.

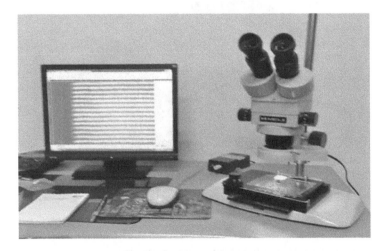

Figure 7.14 Optical microscope to examine the layer thickness.

roughness tester shown in Figure 7.15 was used to measure the roughness of the samples and to achieve the maximum accuracy and consistency. The average surface roughness values on the printed surface were measured perpendicularly to the feed marks at a minimum of fifteen location points of two sides of each sample as shown in Figure 7.16, and five values close to each other were taken as the average surface roughness values by using a scatter chart.

Figure 7.15 Mitutoyo surface roughness tester.

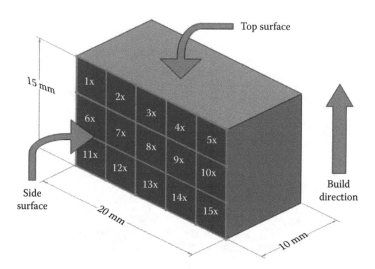

Figure 7.16 Surface area of the sample being measured.

7.4   *Results and discussion*

Ra or the arithmetical means surface roughness is the standard used to evaluate the surface texture. The average surface roughness values on the printed surface were measured perpendicularly to the feed marks at a minimum of fifteen location points of two sides of each sample. A surface roughness tester was further used to measure the roughness of the work-piece. Before conducting the experiments, the calibration of the equipment was done using the standard block in order to ensure consistency. The first sample that was printed with no vibration showed a poor surface finish. The results showed imperfections and discontinuity in the surface, and the range of layer thickness was between 0.17 and 0.25 mm. For the second sample, when an 11 kHz frequency was applied, the results showed that the layer of surface was better than the first piece without vibration being applied. The range of the layer thickness was between 0.07 and 0.09 mm while surface defects were 0.18 and 0.36 mm. For the third sample, when a 16 kHz frequency was applied, the results showed that the defects were lesser compared to that of 11 kHz frequency, and the defect on the surface was reduced to 0.16 mm. However, the thickness between the layers was consistently from 0.08 to 0.09 mm. For the next sample, when a 21, 27 and 40 kHz frequency was applied to the test piece, the result was similar to that of 16 kHz. However, there were less surface defects found. The range of the layer thickness was between 0.07 and 0.08 mm, and the surface roughness was reduced from 14.416 to 13.704 μm. Lastly, when a 50 kHz frequency was applied, the result showed that the thickness between the layers was found to be more compressed, between 0.06 and 0.07 mm thickness, showing even finer layers being produced. Figure 7.17 shows images of the surface roughness of various samples under a microscopic examination.

From these experiments, it was found that the 50 kHz frequency range had produced seams that were much closer together and provided evidence of better surface finish when compared to other samples. As shown from Table 7.4 and Figure 7.18, we can observe that there is an improvement in terms of surface roughness between 21 and 40 kHz with the best results achieved at 50 kHz. This trend is supported by work from Marcel et al. [32] who carried out similar experiments and where the use of a 21 kHz frequency had reduced the surface roughness of parts significantly.

This research has presented a novel approach of an ultrasonic-assisted FDM process to improve surface finish of FDM parts. Three conceptual designs were proposed for the ultrasonic-assisted FDM. Concept 3 was found to be the most feasible due to the safety and reliability aspects of the location of the ultrasonic piezoelectric transducer and 3D printer. From the experiments, a comparative study of the samples being produced showed that ultrasonic vibration did contribute to an improvement of the surface finish of the samples produced. The results showed that a 50 kHz frequency applied to the FDM piece had the best surface finish. Compression had occurred during the built process when ultrasonic vibration was used, in which the layer thickness was reduced to between 0.06 and 0.07 mm. Overall, through the use of ultrasonic vibration, the layer thickness had been reduced from between 0.17 mm to 0.25 mm to 0.06 mm and 0.07 mm, which shows a reduced stair-stepping effect and with a positive improvement over the surface finish.

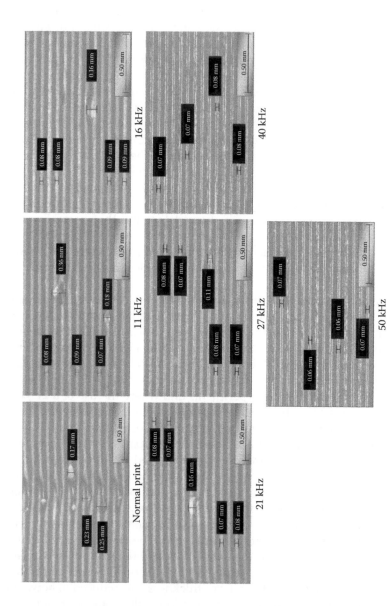

Figure 7.17 Surface roughness of various samples under microscopic examination.

Table 7.4 Data regarding the layer thickness and surface roughness
based on the frequency of oscillation

Piezoelectric type	Frequency, kHz	Layer thickness, mm	Surface roughness, μm
One transducer	0	0.17–0.25	14.360
	11	0.07–0.09	14.374
	16	0.08–0.09	14.284
	21	0.07–0.08	14.416
	27	0.07–0.08	14.032
	40	0.07–0.08	13.704
	50	0.06–0.07	13.658

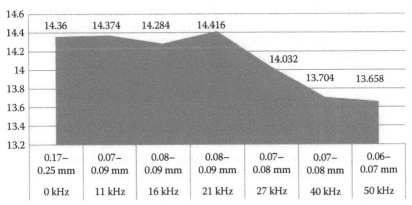

Figure 7.18 Data corresponding to the layer thickness, surface roughness and frequency of piezoelectric oscillation.

The results from this study have potential to be applied to other AM processes such as selective laser sintering, electron beam melting and stereolithography systems. The data from this research has the potential to benefit manufacturers, engineers and those in new product development within the automotive, consumer, medical and sports industries to produce high-quality end-use parts. The main limitation of this study is that it only covers a sample test piece and the experiments are subjected to only the use of ABS material. Future work should be carried out to investigate the application of high frequencies from an ultrasonic transducer to larger and geometrically complex parts to ascertain conclusive findings. We also recommend that future work should involve the use of other methods of 3D surface measurement such as a surface texture gauge [33] or a 3D optical profilometry device. Other materials such as polyactic acid (PLA) or composites should also be tested. Process parameters, such as the layer thickness, build density and the speed of build could be investigated to understand how those parameters could also help improve the quality of the surface finish when used in conjunction with an ultrasonic finishing. More extensive microscopic studies could also help to extend existing knowledge on how ultrasonic finishing could positively or negatively influence the bond between the rasters. For example, Sood et al. [4] claimed that small air gaps help to create strong bonds between two rasters and improve the strength, but small air gaps restrict heat dissipation that give rise to an increased chance of stress accumulation. Other developments could also focus on the impact of ultrasonic vibration to understand the limits of compressive failure for FDM processes.

Acknowledgements

This research was fully sponsored by Universiti Teknikal Malaysia Melaka (UTeM) under the funding of the Fundamental Research Grant Scheme (FRGS) in Malaysia.

Theoretical questions

Q1. Name the different methods of conventional surface finish treatments.
Answer: Product surface finish can be improved through methods such as spray painting, sandblasting, polishing and anodizing.

Q2. How can the quality of the surface of FDM parts be improved?
Answer: FDM parts can be improved through methods such as optimizing the build orientation; slicing strategy (layer thickness); optimizing the fabrication parameters; and through post-treatment.

Q3. How does the proposed method of using ultrasonic vibration for FDM parts work and what level of effectiveness was achieved from this study?
Answer: Ultrasonic vibration induces compression of the layer-by-layer manufacturing process, in which the layer thickness was reduced to between 0.06 and 0.07 mm.

References

1. Wohlers, T. (2016) *Additive Manufacturing and 3D Printing State of the Industry.* Wohlers Associates, Fort Collins, CO.
2. ISO - International Organization for Standardization (2015) *Standard Practice Guide for Design for Additive Manufacturing - ISO 20195:2015.* ISO, Geneva, Switzerland.

3. Galantucci, L.M., Lavecchia, F. and Percoco, G. (2010) Quantitative analysis of a chemical treatment to reduce roughness of parts fabricated using fused deposition modelling. *CIRP Annals-Manufacturing Technology*, 59(1), 247–250.

4. Sood, A.K., Ohdar, R.K. and Mahapatra, S.S. (2012) Experimental investigation and empirical modeling of FDM process for compressive strength improvement. *Journal of Advanced Research*, 3(1), 81–90.

5. Nad, M. (2010) Ultrasonic horn design for ultrasonic machining technologies. *Applied and Computational Mechanics*, 4(1), 79–88.

6. Bual, G.S. and Kumar, P. (2014) Methods to improve surface finish of parts produced by fused deposition modeling. *Manufacturing Science and Technology*, 2, 51–55.

7. Hur, J. and Lee, K. (1998) The development of a CAD environment to determine the preferred build-up direction for layered manufacturing. *International Journal of Advanced Manufacturing Technology*, 14(4), 247–254.

8. Pham, D.T., Demov, S.S. and Gault, R.S. (1999) Part orientation in stereolithography. *International Journal of Advanced Manufacturing Technology*, 15(9), 674–682.

9. Frank, D. and Fadel, G. (1995) Expert system-based selection of the preferred direction of build for rapid prototyping processes. *Journal of Intelligent Manufacturing*, 6(5), 339–345.

10. Phatak, A.M. and Pande, S.S. (2012) Optimum part orientation in rapid prototyping using genetic algorithm. *Journal of Manufacturing Systems*, 31(4), 395–402.

11. Azanizawati, M. (2003) Quality assessment of hollow rapid prototyping model. Master's thesis, Universiti Teknologi, Malaysia.

12. Ahn, D., Kweon, J.H., Kwon, S., Song, J. and Lee, S. (2009) Representation of surface roughness in fused deposition modelling. *Journal of Materials Processing Technology*, 209(15–16), 5593–5600.

13. Kantaros, A. and Karalekas, D. (2013) Fiber Bragg grating based investigation of residual strains in ABS parts fabricated by fused deposition modeling process. *Materials & Design*, 50, 44–50.

14. Boschetto, A., Bottini, L. and Veniali, F. (2016) Finishing of fused deposition modeling parts by CNC machining. *Robotics and Computer-Integrated Manufacturing*, 41, 92–101.

15. Pandey, P.M., Reddy, N.V. and Dhande, S.G. (2003) Improvement of surface finish by staircase machining in fused deposition modelling. *Journal of Materials Processing Technology*, 132(1–3), 323–331.

16. McCullough, E.J. and Yadavalli, V.K. (2013) Surface modification of fused deposition modeling ABS to enable rapid prototyping of biomedical microdevices. *Journal of Materials Processing Technology*, 213(6), 947–954.

17. Espalin, D., Medina, F. and Wicker, R. (2009) Vapor smoothing, a method for improving FDM-manufactured part surface finish, Internal Report of the W.M. Keck Center for 3D Innovation, University of Texas, El Paso, TX.

18. Rao, A.S., Dharap, M.A., Venkatesh, J.V.L. and Ojha, D. (2012) Investigation of post processing techniques to reduce the surface roughness of fused deposition modeled parts. *International Journal of Mechanical Engineering and Technology*, 3(3), 531–544.

19. Boschetto, A. and Bottini, L. (2015) Surface improvement of fused deposition modeling parts by barrel finishing. *Rapid Prototyping Journal*, 21(6), 686–696.

20. Yang, J.J., Zhang, H., Deng, X.Z. and Wei, B.Y. (2013) Ultrasonic lapping of hypoid gear: System design and experiments. *Mechanism and Machine Theory,* 65, 71–78.
21. Kang, B., Kim, G.W., Yang, M., Cho, S.H. and Park, J.K. (2012) A study on the effect of ultrasonic vibration in nanosecond laser machining. *Optics and Lasers in Engineering,* 50(12), 1817–1822.
22. Lian, H., Guo, Z., Huang, Z., Tang, Y. and Song, J. (2013) Experimental research of Al6061 on ultrasonic vibration assisted micro-milling. *Procedia CIRP,* 6, 561–564.
23. Kim, W.J., Lu, F., Cho, S.H., Park, J.K. and Lee, M.G. (2011) Design and optimization of ultrasonic vibration mechanism using PZT for precision laser machining. *Physics Procedia,* 19, 258–264.
24. Tabatabaei, S.M.K., Behbahani, S. and Mirian, S.M. (2013) Analysis of ultrasonic assisted machining (UAM) on regenerative chatter in turning. *Journal of Materials Processing Technology,* 213(3), 418–425.
25. Wang, J., Shimada, K., Mizutani, M. and Kuriyagawa, T. (2013) Material removal during ultrasonic machining using smoothed particle hydrodynamics. *International Journal of Automation Technology,* 7(6), 614–620.
26. Friel, R.J. and Harris, R.A. (2013) Ultrasonic additive manufacturing–A hybrid production process for novel functional products. *Procedia CIRP,* 6, 35–40.
27. Nik, M.G., Movahhedy, M.R. and Akbari, J. (2012) Ultrasonic-assisted grinding of Ti6Al4V alloy. *Procedia CIRP,* 1, 353–358.
28. Janaki Ram, G.D., Robinson, C., Yang, Y. and Stucker, B.E. (2007) Use of ultrasonic consolidation for fabrication of multi-material structures. *Rapid Prototyping Journal,* 13(4), 226–235.
29. Presz, W. and Andersen, B. (2007) Flexible tooling for vibration-assisted microforming. *Journal of Achievements in Materials and Manufacturing Engineering,* 21(2), 61–64.
30. Armillotta, A. (2006) Assessment of surface quality on textured FDM prototypes. *Rapid Prototyping Journal,* 12(1), 35–41.
31. Mousavi, S.A.A., Feizi, H. and Madoliat, R. (2007) Investigations on the effects of ultrasonic vibrations in the extrusion process. *Journal of Materials Processing Technology,* 187–188, 657–661.
32. Marcel, K., Marek, Z. and Jozef, P. (2014) Investigation of ultrasonic assisted milling of aluminum alloy AlMg4.5Mn. *Procedia Engineering,* 69, 1048–1053.
33. Peiponen, K.E., Myllyla, R. and Priezzhev, A.V. (2009) *Optical Measurement Techniques: Innovations for Industry and the Life Sciences,* Vol. 136. Springer, Berlin, Germany, p. 41.

Index

Note: Page numbers in italic and bold refer to figures and tables respectively.

Milton Keynes UK
Ingram Content Group UK Ltd.
UKHW040110071024
449327UK00019B/952